Springer Tracts in Natural Philosophy

Volume 6

Studies in Non-Linear Stability Theory

Wiktor Eckhaus

Springer-Verlag · Berlin · Heidelberg · New York 1965

Wiktor Eckhaus
Département de Mécanique
Université de Paris

Library of Congress Catalog Card Number 65 — 22097

Printed in Germany

Titel No. 6734

Preface

Non-linear stability problems formulated in terms of non-linear partial differential equations have only recently begun to attract attention and it will probably take some time before our understanding of those problems reaches some degree of maturity. The passage from the more classical linear analysis to a non-linear analysis increases the mathematical complexity of the stability theory to a point where it may become discouraging, while some of the more usual mathematical methods lose their applicability. Although considerable progress has been made in recent years, notably in the field of fluid mechanics, much still remains to be done before a more permanent outline of the subject can be established.

I have not tried to present in this monograph an account of what has been accomplished, since the rapidly changing features of the field make the periodical literature a more appropriate place for such a review. The aim of this book is to present one particular line of research, originally developed in a series of papers published in 'Journal de Mécanique' 1962—1963, in which I attempted to construct a mathematical theory for certain classes of non-linear stability problems, and to gain some understanding of the non-linear phenomena which are involved. The opportunity to collect the material in this volume has permitted a more coherent presentation, while various points of the analysis have been developed in greater detail. I hope that a more unified form of the theory has thus been achieved. The discussion of the existing literature has, in this context, been restricted to papers bearing directly on the subjects which are treated. However, a more extensive bibliography of research on non-linear stability problems has also been included.

The theory presented in this monograph is based essentially on two concepts: asymptotic expansions with respect to suitably defined small parameters and series expansions in terms of eigenfunctions. Although asymptotic approximations, in one form or another, have widely been used in the study of non-linear stability problems, the correct formulation and justification of the asymptotic method requires careful analysis, which I have attempted to provide. On the other hand, the concept of eigenfunction expansion seems not to have been recognized until now as a fundamental tool in the analysis of problems of the type considered here; it is this concept which appears to contribute most significantly to the clarification of the mathematical aspects of non-linear behaviour. Although the generalized eigenfunction expansions needed in the analysis still present certain difficulties, which the reader will find discussed as they arise, the concept seems a most promising element for future developments of the non-linear stability theory.

It is a pleasure to record my gratitude to Professor LEON TRILLING, who read the manuscript and suggested innumerable improvements of the text, and to Professor PAUL GERMAIN, whose continuous interest has stimulated the investigations reported here and who invited me to write this monograph.

Paris, April 1965 WIKTOR ECKHAUS

Contents

Chapter 1

INTRODUCTION

1.1 The notion of stability

The concept of stability is so familiar that any introduction may seem superfluous. Yet, on closer analysis, some discussion of the basic ideas appears necessary for the proper formulation of certain problems, in particular the non-linear ones.

The question of stability may, in simple terms, be stated as follows: Given an equilibrium state of a physical system, whose stability we wish to study, we consider a state near equilibrium and ask whether in the course of time the system will tend toward the given equilibrium state.

In somewhat more precise terminology, we shall call the departure from equilibrium a perturbation. The initial value of the perturbation is considered to be given in a problem of stability and the question is asked whether, starting from this initial value, the perturbation will tend to zero in the course of time. If this is the case the equilibrium state is called stable, it is called unstable otherwise.[1]

The mathematical formulation of the stability theory proceeds from the non-linear partial differential equations (or integro-differential equations) which describe the problem of mathematical physics under consideration, under the most general conditions. The unknown quantities are functions of three space coordinates and time, and are subject to some boundary conditions.

Certain special solutions of such a general problem are usually of particular interest; these are the solutions of permanent type (equilibrium states in previous terminology). The simplest among permanent-type solutions are the stationary solutions and we shall restrict our discussion to problems where such solutions exist. In certain cases we may, without worrying about unicity, chose for further study a very particular stationary solution, which depends upon only one or

[1] This definition of 'stability' and 'instability' will be employed throughout in the sequel. It should be noted however that terminology in the field of stability may vary from one author to another. The present definition agrees with LIN (1955) but disagrees for example with LA SALLE and LEFSCHETZ (1961).

two of the three space variables; occasionally even the trivial solution may be of interest. We shall call any of these stationary solutions a basic solution of the problem.

The stability problem for a basic solution is formulated by the following procedure: We suppose that at some initial instant the solution of the general problem consists of the basic solution plus a perturbation, while boundary conditions are not being perturbed. From the equations of the problem we obtain by substitution the equations governing the perturbation quantities. These are again partial non-linear differential equations (or integro-differential equations) for unknown quantities, which are functions of three space variables and time. The equations are homogeneous and are subject to homogeneous boundary conditions.

In the mathematical formulation of the stability problem the initial values of the perturbations are considered given, that is, they can be chosen at will. The question arises now, how should these initial values be chosen; that is, with respect to what perturbations should the stability be studied. To answer that question it becomes necessary to define more precisely the interpretation of the mathematical stability problem in terms of the physical mechanism under consideration.

In appearance the stability problem describes phenomena which may be realized in the physical world. However, this appearance is somewhat misleading and in an attempt to imagine an experiment that would correspond strictly to a problem of stability one faces some conceptiual difficulties. The basic solution (equilibrium state) whose stability is being studied, has no other a priori significance than that of a theoretical abstraction: it represents a possible functioning of the physical system in an idealized world. In studying the stability of the solution in question we admit in fact that conditions for its validity cannot entirely be realized in an experimental investigation. From this point of view the stability theory appears as a confrontation of the theoretician with the irregularities, imperfections and disturbances which are always present in the physical world.

Consider for example a problem of fluid mechanics concerned with the flow in a channel. The stationary laminar solution supposes perfectly smooth walls, perfect regularity of the machinery which sets the fluid in motion and absence of any sources of disturbances external to the apparatus. Needless to say, these conditions cannot be realized. We must therefore first analyse the stability of the solution with respect to perturbations whose nature and level of intensity may be considered representative for the fluctuations that are unavoidable in the experimental apparatus. If we find the solution to be stable, then the theory tells us that every perturbation, after being introduced, decays and tends to zero. In the actual apparatus perturbations are introduced not once, but continuously or intermittently throughout the time-history of the experiment. Thus the theoretical stability problem is of the nature of a stability test, from which we may conclude that the solution under consideration, although it will never be realized, does represent a good approximation for the functioning of the physical system. The amplitudes of the departures from the solution that will be observed in the experiment are determined by the level of fluctuations introduced in the apparatus and can thus presumably be reduced to an acceptable minimum.

Only in the case of stability with respect to the unavoidable level of perturbations can an experiment be performed corresponding to what is imagined in a stability problem: A perturbation of a given intensity and character can be created at a given moment and its evolution in time can be studied. In the case of instability with respect to the unavoidable level no experiment corresponding to the stability problem can be performed, since the initial departure from equilibrium can not be controled (exception being made for some extreemly simple systems, such as a pendulum held by hand in the neighbourhood of its unstable equilibrium). In fact, in the case of instability, the theory shows us that the solution under consideration has no chance of survival in the physical world and that the physical system will function according to some other solution. This *conclusion* is what can be verified in experiment.

Often the theory may predict that with respect to a given level of perturbations the solution is stable if a certain parameter, which can be varied at will, is within a certain range, and unstable otherwise. The boundary between the stable and unstable range is given by a critical value of the parameter and a confirmation of this critical value may be sought by an experimental investigation, which again must start in the stable range and consist of a slow variation of the parameter toward its critical value. In this, of course, care must be taken to maintain the perturbations at the level for which the theory has been established.

If the stability theory is indeed to account for the fluctuations which occur in the physical world, than it must be formulated in such way that the results can be interpreted from the point of view of some realizable experiment. This requirement means that perturbations must in principle be considered as entirely arbitrary (although suitably continuous) functions of the space coordinates. Moreover, it must be admitted that an arbitrary number of different perturbations can be introduced in an arbitrary order of succession an arbitrary number of times during the time-history of the phenomena. The study of a single perturbation as an initial value problem should be considered to be of the nature of a stability test of which the results must be interpreted with care. This is particularly pertinent in non-linear problems where, in certain cases of which we shall see later an example, successive introduction of two different perturbations may result in a different behaviour than what can be inferred from the separate study of the two perturbations.

On the other hand, we do have the liberty of fixing at will the level of perturbations (that is, their intensity defined in some suitable way) with respect to which we wish to study the stability. This reflects the assumption that by refining the experimental apparatus the 'unavoidable fluctuations' may be made smaller and smaller, while perturbations of any given intensity may be created by artificial means. In a limiting process we can even study perturbations of which the level is arbitrarily small, although it can never quite equal zero.

1.2 The nature of non-linear stability problems

The classical theory of stability is the linearized theory: we assume the perturbations arbitrarily small and in the equations and boundary conditions of the problem we neglect terms non-linear in the perturbation-quantities as compared to the linear ones. We then study the stability of the basic solution with respect to perturbations of infinitesimally low level. Obviously such perturbations are unavoidable in any experimental apparatus.

If now the physical system under consideration is 'autonomous', —that is, if the time variable does not appear explicitly in the equations and boundary conditions—, and if the basic solution is stationary, then the linearized equations of the stability problem admit solutions of which the behaviour with time is given by an exponential function. Instability is characterized by the existence of perturbations which grow without limit in the course of time; in the case of stability all perturbations tend to zero as time tends to infinity. A limiting case between stability and instability, which we shall call 'neutral stability', occurs when there exists a perturbation which, after being introduced, remains of essentially constant amplitude in the course of time, while all other possible perturbations tend to zero.

As an illustration let us consider the following simple problem which arises in connection with BURGERS' mathematical model of turbulence to be studied in some detail in chapter 5. Suppose that the function $\Phi(\eta, t)$ is a perturbation which in the linearized theory satisfies the equation

$$\frac{1}{R}\frac{\partial^2 \Phi}{\partial \eta^2} + \Phi - \frac{\partial \Phi}{\partial t} = 0; \qquad 0 \leqq \eta \leqq 1 \tag{1.2.1}$$

and boundary conditions

$$\Phi(0, t) = \Phi(1, t) = 0 \tag{1.2.2}$$

R is a parameter which may be varied at will.

The solutions of the problem are given by

$$\Phi_n = A_n\, e^{-\mu_n t} \sin(n+1)\,\pi\,\eta; \qquad n = 0, 1, 2, \ldots \tag{1.2.3}$$

where A_n are arbitrary constants to be determined by the initial conditions, and where μ_n is given by

$$\mu_n = -\left[1 - \frac{\pi^2}{R}(n+1)^2\right] \tag{1.2.4}$$

There exists thus a denumerable infinite sequence of possible perturbations Φ_n. If $R < \pi^2$ all perturbations tend to zero and we have stability; if $R > \pi^2$ one or more perturbation increases without limits and we have instability. The limiting case of neutral stability occurs for $R = \pi^2$. In that case Φ_0 remains constant in the course of time, while Φ_n for $n > 0$ tend to zero as $t \to \infty$.

We remark that in the linearized stability theory the superposition principle holds. Consequently, it makes no difference how many times and in what order of succession different perturbations have been introduced.

The linearized analysis may be considered as the first step in any stability theory; moreover, it is a natural starting point for the description and definition of non-linear problems. Obviously, non-linear stability problems arise whenever linearized theory is no longer adequate and non-linear terms of the equations have to be taken into account.

Let us consider first the case of stability in the sense of the linearized theory. We know that in this case infinitesimal disturbances decay to zero. Now infinitesimal disturbances are certainly unavoidable, but not all unavoidable disturbances may be considered infinitesimal. In other words, for the explanation of some physical phenomena we may be led to study the stability with respect to perturbations whose level can no longer be assumed infinitesimally small. The answer to the stability question depends then on the level of perturbation. We may expect that under certain conditions, while there is stability with respect to extremely small perturbations, an instability arises with respect to perturbations of level higher than a critical value. This behaviour, called 'instability to finite size perturbations', will in fact present itself frequently in the course of our study and may be considered one of the most typical phenomena associated with non-linear stability problems.

Consider next the case of instability with respect to infinitesimal perturbations. There is, in this case, an apparent contradiction within the linearized theory, since we assume very small perturbations and find that they grow without bounds. From the mathematical point of view the paradox may be circumvented by stating that we consider infinitesimal initial values and concluding, when we find that perturbations grow without limits, that the solution is only valid during some finite initial time interval. This naturally poses the question: what happens to the perturbations when they have grown so large that linearization can no longer be considered valid.

The behaviour of perturbations in the case of instability is a problem of fundamental interest and various types of behaviour are conceivable. We shall particularly be interested in the question of existence of non-linear bounds on the growth of perturbations. Thus, in studying the non-linear problem we may find that the amplitude of perturbation, instead of growing without limits, tends to a finite value as time tends to infinity. Moreover, this finite value may be small, although not infinitesimal. If that is the case, the basic solution under consideration, although unstable in the ordinary sense, still represents a good approximation of the functioning of the physical system. In constrast to the stable case, however, the level of departures from the solution observed in an experimental investigation will not be bound by the intensity of externally introduced fluctuations, but will be, so to speak, a property of the system itself.

The non-linear problems arising in cases of stability and of instability described above are in many respects complementary and we shall find that they may often be treated in one and the same analysis. The essential mathematical difficulty associated with these problems is illustrated by considering for example the case of instability.

If we assume that initial values of the perturbations are infinitesimally small, then the linearized theory may be considered an approximation valid during some finite initial time interval. This suggests that we may try to construct the solution of the non-linear problem by an iteration procedure, in which the linearized solution is taken as the first approximation. We would then obtain a series solution, every term of the series growing without limit as a function of time.[1] The convergence of the series can of course be investigated, although mathematical complexity may make such analysis prohibitive. Most certainly the series will not converge for all times and in particular for every fixed initial value, no matter how small, the series will not converge in the limit as time tends to infinity. On the other hand, it is this behaviour in the limit that we are interested in, since we want to find out whether the perturbation approaches a finite value. The nature of the problem

[1] For a recent example of this approach see for instance STRUMINSKY (1963a).

thus requires that, no matter what technique of approximation is used, certain essential non-linear effects be included in the first approximation.

The above discussion can be illustrated by a simple example: Consider the equation

$$\frac{dA}{dt} = A\left(1 - \frac{A^2}{\alpha^2}\right)\left(1 + \frac{A^2}{\beta^2}\right) \tag{1.2.5}$$

where α and β are constants, $|\alpha| \ll |\beta|$, and where a real initial value $A(0) = a_0$ is given, such that $|a_0| \ll 1$. In a first approach we can, by an iteration procedure, construct a series solution in which the linearized solution is the first approximation and obtain a series of which every term grows without limit as a function of time. In fact, we have defined by this procedure an expansion valid in the vicinity of the singular point $A = 0$ of the equation. There is however another singular point, $A = \alpha$, which is of greater interest from the present point of view. What we wish is an approximation valid in a region containing both $A = 0$ and $A = \alpha$. We can obtain the desired result by an iteration procedure in which, as first approximation, we take the solution of the equation

$$\frac{dA}{dt} = A\left(1 - \frac{A^2}{\alpha^2}\right) \tag{1.2.6}$$

This leads to a series solution which can be shown to be convergent. The solution tends in the limit $t \to \infty$ to the singular point $A = \alpha$; inclusion of more terms in the series modifies the manner in which this point is approached.

In problems of greater complexity, when the unknown quantity is a function of space variables and time, the aim of the mathematical analysis is still essentially the same as in the simple problem above. We wish to construct a solution of which the region of validity includes vanishingly small perturbations as well as the equivalent of the singular point $A = \alpha$. However, in general we do not know a priori where this 'singular point' lies, nor are we even sure that it exists.

It should be clear now that non-linear problems of stability require special mathematical techniques. Before discussing these, however, it is useful to illustrate the nature of non-linear phenomena in stability from yet another point of view: that of experimental evidence. We shall briefly describe certain types of non-linear behaviour observed in experimental investigations in the field of fluid mechanics.

Perhaps the most classical problem of hydrodynamic stability is that of the flow between two coaxial rotating cylinders. First investigated theoretically and experimentally by G. I. TAYLOR (1923), it has been the subject of numerous studies ever since (see, for example, LIN (1955)). The predictions of the linearized theory have been confirmed by experiments. Moreover, in the case of instability it was found that under certain conditions a regular and time-independent pattern of secondary flow is established. This secondary flow can, no doubt, be interpreted as a perturbation, which in the linearized theory grows without limits, but which here, due to non-linear effects, has reached a final stationary amplitude.

A mathematically similar problem, with similar behaviour, arises in connection with the convective instability of a fluid layer heated from below. Experimental and theoretical investigations have been originated by BENARD (1901) and by RAYLEIGH (1916), recent developments of the linearized theory may be found in CHANDRASEKHAR (1961). Here again, one observes a reasonably regular pattern of secondary flow, which usually has the appearence of hexagonal cells and which may be interpreted as a manifestation of a non-linear limit to the growth of perturbations.

Cellular structure of more oscillatory character occurs in flame instability. Experimental evidence in this domain is provided by MARKSTEIN's (1951) investigations; theoretical studies have been orginated by LANDAU (see LANDAU and LIFSCHITS), further developments can be found in ECKHAUS (1961).

In all these cases, the phenomena ultimately observed in experiment still resemble the unstable perturbation predicted by linearized theory, only the amplitude is limited in size. It is for this reason that the behaviour may be interpreted in terms of a rather simple non-linear effect, that of a non-linear bound on the growth of perturbations. The situation changes radically when turbulence makes its appearence.

SCHUBAUER and SKRAMSTAD's (1947) famous experiments on boundary layer instability have revealed that unstable perturbations do appear as predicted by the linearized theory and grow for some time (or, more exactly, over some distance) roughly in accordance with the theory. Further development is however suddenly broken off by the appearance of a new phenomenon: the transition to turbulent motion. In other unstable flows turbulence may make its appearence in somewhat different and less explosive fashion (see for example SATO and KURIKI's (1961) experiments on wake-flow), but in all cases it is clear that the simple mechanism of a non-linear bound on the growth of linearly unstable perturbations is irrelevant to the essential phenomenon. On the contrary, the occurrence of turbulence is more in the nature of a new instability, of essentially non-linear character.

Admitting that turbulence occurs through some kind of instability, in certain cases of linearly stable flows we have examples of behaviour that we have called instability to finite size perturbations. Thus, in the experiments of SCHUBAUER and KLEBANOFF (1956) turbulence was triggered by excitations of sufficient intensity in a region of boundary-layer flow that is stable according to linearized theory. On the other hand, REYNOLDS' (1883) experiments, which lay at the very origin of the study of instability and turbulence, were concerned with flow in circular pipes. while linearized stability theory has not revealed an instability for fully developed pipe-flow. Yet turbulence has been observed under various circumstances and various explanations have been advanced (notably in terms of instability of the boundary layer in the inlet region). From the point of view of non-linear stability theory it should be noted that in modern investigations it was found possible to keep the flow laminar at higher and higher Reynolds-numbers by reducing the level of 'inevitable perturbations' to lower and lower values. In this connection we mention the experimental studies of LEITE (1959), whose conclusion was that '. . . transition to turbulent flow occurs whenever the amplitude of the disturbances exceeds a threshold value which increases with decreasing Reynolds-number . . .'

Theoretical research on non-linear stability problems in fluid mechanics has been pursued actively in recent years and has resulted in considerable progress. For an excellent account of what has been achieved the reader is referred to STUART's (1960a) review; more recent publications are included in the Bibliography. We shall have the opportunity to discuss briefly some of this work in chapter 9, in connection with the problem of Poiseuille flow.

1.3 Formal approach to stability theory

As we have seen in section 1.1, the theory of stability is concerned with homogeneous partial non-linear differential equations with homogeneous boundary conditions imposed on some limiting surfaces. In studying the non-linear problems in particular, our goal is to develop satisfactory methods of solution and analyse the behaviour of perturbations. But if we are concerned with methods rather then with quantitative results there is no need to confine the attention to a particular problem. On the contrary, we can study classes of mathematical problems without apparent links with physical systems, provided that these problems are chosen so as to be representative of physical stability problems of interest, and provided secondly that in the choice of perturbations we respect the considerations of section 1.1. This is the approach adopted in the present study. It was originally developed by this author in a series of papers (1962—1963) published in *Journal de Mécanique,* which have served as a basis for the chapters that follow.

The formal approach to stability theory has the advantage of greater generality and flexibility. Thus, within certain limits, we may chose the problems at will and follow a course of study which seems natural from the point of view of mathematical analysis: we start with simple problems and with the understanding and experience gained in their study proceed to more complicated ones.

It is with these considerations in mind that we shall first consider (chapters 2 to 5) a class of problems in one dimensional space. Those problems do not claim to reflect the functioning of any physical system, but as mathematical models they represent certain features of non-linear stability theory in its simplest form.

Our analysis will be based mainly on asymptotic developments with respect to suitably defined small parameters, and on expansions in series of eigenfunctions of the linearized theory. When we proceed to more realistic problems in two-dimensional space (chapters 6 to 8) we shall find the same mathematical tools useful, although in somewhat modified form. Indeed, the eigenfunction expansion in particular can be said to be characteristic of the method of analysis developed and utilised throughout this study.

In a final chapter we shall consider an application of the formal theory to the problem of stability of Poiseuille flow. This will allow us to confront our analysis with results obtained earlier by different methods.

A word should be said about the choice of the classes of problems studied below. We shall assume that boundary conditions are imposed on surfaces that are fixed in space. This leaves out of consideration problems associated with so-called 'free-boundaries', which, in this authors opinion, require a different approach. To simplify the equations we shall consider problems involving one unknown quantity and thus governed by a single non-linear partial differential equation. Furthermore, the equation to be considered will involve only a first order derivative with respect to time, which will appear linearly. Thus the problems to be studied are governed by non-linear 'diffusion-type' equations.

Meaningful stability problems associated with diffusion-type equations arise for instance in fluid mechanics, but it is not only the authors familiarity with this field which has led to that choice. In fact, the study of diffusion-type equations appears quite natural if one is interested in the stability of *stationary* solutions.

On the other hand, in most cases in which the governing equation contains higher order derivatives with respect to time, problems of stability and oscillations are interwoven and one is led to study at the outset the stability of *non-stationary* solutions of permanent type.

In the present study we shall eventually consider certain oscillatory permanent-type solutions and their stability, but these solutions will arise as by-products of the stability-analysis of stationary solutions, which will be our major concern.

Chapter 2

A CLASS OF PROBLEMS IN ONE-DIMENSIONAL SPACE

2.1 Preliminary remarks

In the present chapter we begin the study of a class of mathematical problems which, though relatively simple, may still be interpreted as non-linear problems of stability. These problems do not necessarily reflect any physical reality; they represent mathematical models helpful in gaining understanding of certain non-linear phenomena. Moreover, from the mathematical point of view, the problems to be considered already contain certain difficulties which are typical of non-linear stability theory, so that their study can be considered as a first step in the construction of adequate methods of analysis for problems of greater complexity.

The fundamental simplification that we introduce here is the restriction to one-dimensional space. Even in this restricted domain we shall not seek great generality. Our aim, in these first few chapters, is to study what perhaps can be defined as the class of simplest possible non-linear stability problems.

We shall be concerned with problems governed by non-linear partial differential equations for unknown quantities which are functions of one space-variable and time. We shall naturally assume that solutions exist and, as far as the general time-dependent problem is concerned, we shall assume uniqueness, that is, we shall suppose that to every set of initial values there corresponds one and only one solution (although, as we shall see, certain solutions may correspond in the limit as time tends to infinity to whole classes of initial values). The general time dependent problem is thus supposed to be in a sense a 'properly posed' problem, however, this does not assure the uniqueness of the basic solution (that is, solution of the time-independent problem), nor is the case where the basic solution is unique of any exceptional interest. Indeed, if there exist several basic solutions, some of them will be stable while other will be unstable, and it is the aim of the stability-analysis to discern which solution may be realized, under what conditions.

2.2 Formulation

Let η be the space variable and t the time. The function $\Phi(\eta, t)$ satisfies the non-linear partial differential equation

$$L(\Phi) - \frac{\partial \Phi}{\partial t} = F(\Phi) \tag{2.2.1}$$

where L is a linear operator and $F(\Phi)$ denotes the non-linear terms. We shall assume that L is a real linear differential operator in η, independent of the variable t. Thus $L(\Phi)$ is a homogeneous differential expression of the form

$$L(\Phi) = A\,\Phi + B\,\Phi_\eta + C\,\Phi_{\eta\eta} + \dots \tag{2.2.2}$$

where A, B, C, etc. are real functions of η, we shall assume that $C \neq 0$.

We shall restrict the form of the non-linear terms $F(\Phi)$ as follows: let $\mathfrak{L}_i$ be any real linear operator independent of t; we shall assume that $F(\Phi)$ is a linear combination of terms of the form $\mathfrak{L}_i(\Phi^p)\ \mathfrak{L}_j(\Phi^q)$ of which the degree $p + q$ equals 2 or 3, while p and q are integers. Thus $F(\Phi)$ may contain for example the terms Φ^2, Φ_η^2, $\Phi\Phi_\eta^2$, etc. as well as terms involving integrals with respect to η.

It is seen that time only enters the problem through the term $\partial\Phi/\partial t$ of Eq. (2.2.1), which defines the equation in the 'diffusion-type'.

We consider Eq. (2.2.1) to be valid within the finite interval $0 \leqq \eta \leqq 1$; at the limits $\eta = 0$ and $\eta = 1$ of the interval boundary conditions are imposed which we shall assume to be linear, time-independent, but not necessarily homogeneous. For example, if Eq. (2.2.1) is of second order, we can prescribe at the boundaries the value of Φ or that of its derivative.

We shall assume that there exists a time-independent function $\Phi_0(\eta)$ which satisfies Eq. (2.2.1) and the boundary conditions. Φ_0 is hence a basic solution of the problem and the subject of our study is the stability of this basic solution. We write

$$\Phi(\eta, t) = \Phi_0(\eta) + \Phi'(\eta, t) \tag{2.2.3}$$

where $\Phi'(\eta, t)$ is the perturbation, of which the initial value $\Phi'(\eta, 0)$ is supposed to be known. In other words, we study the stability of the basic solution Φ_0 with respect to a given initial perturbation $\Phi'(\eta, 0)$.

Since Φ_0 satisfies the equation

$$L(\Phi_0) = F(\Phi_0) \tag{2.2.4}$$

and the boundary conditions of the problem, the perturbation $\Phi'(\eta, t)$ must satisfy the equation

$$L(\Phi') - \frac{\partial\Phi'}{\partial t} = F(\Phi_0 + \Phi') - F(\Phi_0) \tag{2.2.5}$$

and homogeneous boundary conditions. The right-hand-side of Eq. (2.2.5) contains some linear and some non-linear terms but by rearranging Eq. (2.2.5) we easily see that it is of the type defined in Eq. (2.2.1).

We note that although both the basic solution and the perturbation satisfy equations belonging to the *class* (2.2.1), when a particular problem is considered, the operator L and the definition of the non-linear terms will be different in the equation satisfied by the basic solution and in that governing the perturbation. More precisely, if Φ satisfies Eq. (2.2.1) and Φ_0 is a solution of Eq. (2.2.4), the perturbation Φ' will be a solution of the equation

$$L'(\Phi') - \frac{\partial \Phi'}{\partial t} = F'(\Phi') \tag{2.2.6}$$

which belongs again to the class defined in (2.2.1).

The stability problem consists thus in solving an equation of the type (2.2.1) with homogeneous boundary conditions and with prescribed initial values.

2.3 Behaviour and properties of the linearized solutions

When considering perturbations that are extremely small, we may feel justified in neglecting the non-linear terms in Eq. (2.2.6) as compared to the linear ones. In the resulting linearized stability theory we then have the equation

$$L'(\Phi') - \frac{\partial \Phi'}{\partial t} = 0 \tag{2.3.1}$$

with homogeneous boundary conditions. Obviously, Eq. (2.3.1) admits solutions of the form

$$\Phi'(\eta, t) = e^{-\mu t} \varphi(\eta) \tag{2.3.2}$$

The function φ must then satisfy the ordinary linear differential equation

$$[L' + \mu]\,\varphi = 0 \tag{2.3.3}$$

and homogeneous boundary conditions.

Non-trivial solutions for φ exist only if the parameter μ takes special values $\mu = \mu_n$, which are the eigenvalues of the linearized problem. The corresponding solutions $\varphi = \varphi_n$ are the eigenfunctions. The main aim of the linearized stability theory is the determination of the eigenvalues μ_n. Evidently, if for all eigenvalues $Re(\mu_n) > 0$, then by Eq. (2.3.2) the perturbations decay with time and the basic solution can be said to be stable with respect to infinitesimal perturbations. If, however, $Re(\mu_n) < 0$ for some eigenvalues, then there is instability and in the linearized theory the unstable perturbations grow without limit as functions of time.

There exists a class of operators L' for which several important general properties of the eigenvalues are well-known. If L' is a second order self-adjoint differential operator, with the coefficients A, B, and C, of Eq. (2.2.2) satisfying certain simple conditions, then the problem for φ is a Sturm-Liouville problem.[1] In that case all eigenvalues are

---

[1] The theory of the Sturm-Liouville problem can be found for instance in MORSE and FESHBACH (1953).

real and the sequence of eigenvalues μ_n is an infinite denumerable sequence. Moreover, the eigenvalues may be ordered in such a way that

$$\mu_{n+1} > \mu_n; \quad n = 0, 1, 2, \ldots \tag{2.3.4}$$

If the problem for φ is not a Sturm-Liouville problem, then several of the properties of eigenvalues mentioned above remain true in general, but they have to be proved for each individual case. In what follows we shall not restrict ourselves to Sturm-Liouville operators; we shall assume however the operator L' to be such that the following properties hold:

I. There exists an infinite denumerable sequence of eigenvalues μ_n
II. The eigenvalues may be ordered in such way that

$$Re[\mu_{n+1}] > Re[\mu_n]; \quad n = 0, 1, 2, \ldots \tag{2.3.5}$$

In chapters 3 and 4 we shall, for reasons of simplicity, consider only cases of real eigenvalues μ_n; in the present chapter however there is no need to make this restriction .Moreover, when considering problems in two-dimensional space (chapter 6 and sequel) we shall have to admit complex eigenvalues and continue, in a sense, the analysis of this chapter.

What we have excluded from further consideration is the possibility of continuous spectra of eigenvalues. Such continuous spectra usually arise in eigenvalue-problems associated with infinite intervals (and this is why we have explicitly chosen to consider the finite interval $0 \leqq \eta \leqq 1$); they arise also if coefficients of the differential operator L' are singular at the boundaries.

We can write now a solution of the linearized stability problem in the form

$$\Phi'(\eta, t) = \sum_{n=0}^{\infty} b_n e^{-\mu_n t} \varphi_n(\eta) \tag{2.3.6}$$

where b_n are constants to be determined by the initial conditions. Implicitely we assume here that the series (2.3.6) converges.

A stability problem usually contains at least one parameter, characteristic of the basic solution under consideration, whose variation may change the answer to the stability question. We shall consequently suppose that the operator L' contains a parameter R, and that there exists a value $R = R_{cr}$ such that $Re(\mu_n) > 0$ for all values of n if $R < R_{cr}$, but $Re(\mu_n) < 0$ for at least $n = 0$ if $R > R_{cr}$. We shall say that the basic solution is stable if $R < R_{cr}$, that it is unstable if $R > R_{cr}$, and that for $R = R_{cr}$ we have neutral stability.

We remark that in accordance with property (2.3.5), if R exceeds R_{cr} slightly, we have $Re(\mu_0) < 0$ but $Re(\mu_n) > 0$ for $n > 0$. For higher values of R we may have $Re(\mu_0) < 0$, $Re(\mu_1) < 0$ but $Re(\mu_n) > 0$ for $n > 1$; etc.

2.4 Series expansion in the case of self-adjoint operators

We shall now assume the linearized stability problem to be completely solved. Suppose, for example, that we have found an instability such that $Re(\mu_0) < 0$ and $Re(\mu_n) > 0$ for $n \geqq 1$. Then in the linearized theory, Eq. (2.3.6), the perturbation is represented by a series of which all terms tend to zero as time passes, except for the first term which tends to infinity. After a certain interval of time this behaviour contradicts the basic assumption of the linearized theory and non-linear effects have to be taken into account. However, if the initial values of the perturbation are sufficiently small, then we may expect the linearized theory to be, in some sense, a valid approximation in some initial time interval and in constructing a non-linear theory we may wish to be able to follow the development of the perturbations whose behaviour is initially given be Eq. (2.3.6), beyond the domain of its validity. In other words, in the non-linear theory we would like to keep a link with the linearized results.

In fact, as we shall see shortly, the link with the linearized theory arises in a natural way and the results of the linearized analysis provide a powerful tool for the study of the non-linear problem.

We start with the case where L' is a Sturm-Liouville operator. The sequence of eigenfunctions φ_n is then a complete set of orthogonal functions in the interval $0 \leqq \eta \leqq 1$. We may suppose these functions normalized, so that

$$\int_0^1 \varphi_n \varphi_m \, d\eta = \begin{cases} 1 & \text{if} \quad n = m \\ 0 & \text{if} \quad n \neq m \end{cases} \tag{2.4.1}$$

We now have the following expansion theorem: Any continuous function with piecewise continuous first and second derivatives, which satisfies the boundary conditions of the linearized stability problem, can be expanded in the interval $0 \leqq \eta \leqq 1$ into an absolutely and uniformly convergent series of eigenfunctions (COURANT-HILBERT Vol. I).

Consequently, the solution $\Phi'(\eta, t)$ of the non-linear stability problem can at any instant of time be represented by the series-expansion

$$\Phi'(\eta, t) = \sum_{m=0}^{\infty} A_m(t)\, \varphi_m(\eta) \tag{2.4.2}$$

where $A_m(t)$ are 'amplitude functions' which have yet to be determined.

Turning now to a more general case of self-adjoint operator L' which is not of the Sturm-Liouville type (for example, a higher order self-adjoint differential operator) we remark that the orthogonality property (2.4.1) of the eigenfunctions is preserved. We have already assumed that the eigenvalues form an infinite denumerable sequence, hence also in this case, under suitable continuity conditions, the expansion formula (2.4.2) holds.[1]

In order to determine the amplitude functions $A_m(t)$ we utilize Eq. (2.2.6) which governs the problem. We introduce the expansion (2.4.2), multiply Eq. (2.2.6) by the eigenfunction φ_n and integrate with respect to η over the interval. We then obtain

$$\frac{d An}{dt} + \mu_n A_n = -\int_0^1 F'(\Phi')\, \varphi_n(\eta)\, d\eta \tag{2.4.3}$$

which holds for any amplitude function $A_n(t)$, $n = 0, 1, 2, \ldots$

We remark, in passing, that in the case of the self-adjoint operator considered here, Eq. (2.4.3) when multiplied by A_n can be interpreted as an equation describing the time evolution of energy contained in the 'component' $A_n(t)\, \varphi_n(\eta)$ of the perturbation.

2.5 Series expansion in the case of not self-adjoint operators

If the operator L' is not self-adjoint, then the eigenfunctions φ_n of the linearized theory are not mutually orthogonal. However, we can define a second infinite denumerable sequence of functions $\tilde{\varphi}_n$, such that φ_n and $\tilde{\varphi}_n$ form a bi-orthogonal set of functions, that is

$$\int_0^1 \varphi_m \tilde{\varphi}_n\, d\eta = \begin{cases} 1 & \text{if} \quad n = m \\ 0 & \text{if} \quad n \neq m \end{cases} \tag{2.5.1}$$

The desired functions $\tilde{\varphi}_n$ are solutions of the eigenvalue-problem that is adjoint to the problem for φ. We define the adjoint operator $\tilde{L}'$ by the relation

$$\psi_1\, L'(\psi_2) - \psi_2\, \tilde{L}'(\psi_1) = \frac{d}{d\eta} P(\psi_1, \psi_2) \tag{2.5.2}$$

where ψ_1 and ψ_2 are any two (suitably continuous) functions of η, and $P(\psi_1, \psi_2)$ is a bi-linear form, often called the bi-linear concomitant. The adjoint problem is now defined by

$$[\tilde{L}' + \lambda]\, \tilde{\varphi} = 0 \tag{2.5.3}$$

with homogeneous boundary conditions chosen such that $P(\varphi, \tilde{\varphi})$ vanishes at $\eta = 0$ and $\eta = 1$. One easily verifies that under these conditions the orthogonality relation (2.5.1) holds (see, for instance, MORSE and FESHBACH).

[1] The theory of eigenfunction expansions in the case of higher order self-adjoint operators has been developed by KODAIRA (1950).

The utility of this bi-orthogonal set of functions is of course subordinate to the requirement that the expansion (2.4.2) be convergent. In other words, we must assume that, as in the case of the Sturm-Liouville problem, an expansion theorem can be established, permitting to define convergent series of φ_n. This being granted, we obtain the equations for the amplitude functions $A_n(t)$ by multiplying Eq. (2.2.6) be the adjoint eigenfunction $\tilde{\varphi}_n(\eta)$ and integrating with respect to η over the interval. The result is

$$\frac{dA_n}{dt} + \mu_n A_n = -\int_0^1 F'(\Phi')\,\tilde{\varphi}_n(\eta)\,d\eta \tag{2.5.4}$$

Eq. (2.5.4) is a more general form of Eq. (2.4.3), and reduces to Eq. (2.4.3) when the operator L' is self-adjoint.

2.6 Interpretation of the series expansion in terms of the Green's function

The eigenfunction expansion considered in the preceeding sections can also be obtained by a somewhat different procedure, based on the concept of Green's function. We shall presently briefly outline this approach.

Returning to Eq. (2.2.6), we can write it in the form

$$L'(\Phi') - \frac{\partial \Phi'}{\partial t} = -\psi(\eta, t); \qquad -\psi(\eta, t) = F'(\Phi') \tag{2.6.1}$$

that is, we can treat the non-linear equation as a non-homogeneous counter-part of the linearized equation.

The solution of the non-homogeneous Eq. (2.6.1), subject to given homogeneous boundary conditions, can be written formally as

$$\Phi'(\eta_0, t_0) = \int^{t_0} \int_0^1 G(\eta, \eta_0; t, t_0)\,\psi(\eta, t)\,d\eta\,dt \tag{2.6.2}$$

where $G(\eta, \eta_0; t, t_0)$ is the Green's function for the linearized problem. If a closed form expression for the Green's function were known, we would obtain by Eq. (2.6.2) an integral-equation formulation of the non-linear problem. In the absence of a closed form, we can still define the Green's function by means of eigenfunctions[1]. In fact, it is easily verified that in the present case the eigenfunction expansion of the Green's function reads

$$G(\eta, \eta_0; t, t_0) = \sum_{n=0}^{\infty} e^{-\mu_n(t_0-t)}\,\varphi_n(\eta_0)\,\tilde{\varphi}_n(\eta) \tag{2.6.3}$$

---

[1] For a detailed discussion of the eigenfunction expansion of Green's function the reader is referred again to Morse and Feshbach chapt. 7 and Courant-Hilbert Vol. I, chapt. V.

Introducing Eq. (2.6.3) into Eq. (2.6.2) we obtain

$$\Phi'(\eta_0, t_0) = \sum_{n=0}^{\infty}\left\{\int_0^{t_0}\int_0^1 e^{-\mu_n(t_0-t)}\,\tilde{\varphi}_n(\eta)\,\psi(\eta, t)\,d\eta\,dt\right\}\varphi_n(\eta_0) \qquad (2.6.4)$$

If we now write

$$A_n(t_0) = \int_0^{t_0}\int_0^1 e^{-\mu_n(t_0-t)}\,\tilde{\varphi}_n(\eta)\,\psi(\eta, t)\,d\eta\,dt \qquad (2.6.5)$$

then expression (2.6.4) becomes identical with the series expansion (2.4.2). On the other hand, one readily shows that the definition of A_n by Eq. (2.6.5) is entirely equivalent to Eq. (2.5.4)

2.7 The system of equations for the amplitude-functions

We now return to Eq. (2.5.4). If the form of the non-linear terms is restricted by the conditions formulated in section 2.2, then the result of the calculation of the integral on the right-hand-side of Eq. (2.5.4) will always be of the form

$$-\int_0^1 F'(\Phi')\,\tilde{\varphi}_n(\eta)\,d\eta = \sum_{ij}\alpha_{ij}^{(n)}\,A_i\,A_j + \sum_{ijr}\beta_{ijr}^{(n)}\,A_i\,A_j\,A_r \qquad (2.7.1)$$

the summation symbols in Eq. (2.7.1) denote double and triple sums respectively, that is:

$$\sum_{ij} = \sum_{i=0}^{\infty}\sum_{j=0}^{\infty}\,; \qquad \sum_{ijr} = \sum_{i=0}^{\infty}\sum_{j=0}^{\infty}\sum_{r=0}^{\infty}$$

Eq. (2.5.3) becomes

$$\frac{dA_n}{dt} + \mu_n A_n = \sum_{ij}\alpha_{ij}^{(n)}\,A_i\,A_j + \sum_{ijr}\beta_{ijr}^{(n)}\,A_i\,A_j\,A_r \qquad (2.7.2)$$

The coefficients $\alpha_{ij}^{(n)}$ and $\beta_{ijr}^{(n)}$ can be evaluated explicitly for every particular form of $F'(\Phi')$. Thus, all coefficients of Eq. (2.7.2) can be considered as known quantities, which can be calculated if the solutions of the linearized problem and its adjoint are available. In what follows we shall, for reasons of simplicity, assume that μ_n, $\alpha_{ij}^{(n)}$ and $\beta_{ijr}^{(n)}$ are all real; for this it is necessary and sufficient that μ_n, φ_n and $\tilde{\varphi}_n$ be real. Since $\Phi'(\eta, t)$ is supposed to be representative for some physical quantity, it has to be real too. Consequently, only real solutions of Eq. (2.7.2) will be considered as meaningful.

Eq. (2.7.2) holds for every value of $n = 0, 1, 2, \ldots$, hence for the time-dependent amplitude-functions A_n we have an infinite system of coupled non-linear first order differential equations. The study of the behaviour of solutions of this system, and the effective construction of these solutions, will be our major concern in the two chapters that follow.

We remark that in the linearized limit, that is if we neglect the right-hand-side of Eqs (2.7.2), we find again the linearized solution for A_n as we should. This is the only case in which every amplitude A_n is independent of all other amplitudes A_m, $n \neq m$. In general, because of non-linear effects, there will be interdependence, which can be brought out more clearly by regrouping the terms of Eqs. (2.7.2). For this purpose we write

$$\frac{dA_n}{dt} = \beta^{(n)} A_n^3 + [\alpha^{(n)} + \psi_2^{(n)}] A_n^2 + [-\mu_n + \psi_1^{(n)}] A_n + \psi_0^{(n)} \quad (2.7.3)$$

where we have abbreviated

$$\beta^{(n)} = \beta^{(n)}_{nnn}; \qquad \alpha^{(n)} = \alpha^{(n)}_{nn} \quad (2.7.4)$$

and where

$$\left.\begin{aligned} \psi_2^{(n)} &= \sum_i \beta^{(n)}_{inn} A_i + \sum_j \beta^{(n)}_{njn} A_j + \sum_r \beta^{(n)}_{nnr} A_r \\ \psi_1^{(n)} &= \sum_i \alpha^{(n)}_{in} A_i + \sum_j \alpha^{(n)}_{nj} A_j + \sum_{ij} \beta^{(n)}_{ijn} A_i A_j + \\ &\quad + \sum_{ir} \beta^{(n)}_{inr} A_i A_r + \sum_{jr} \beta^{(n)}_{njr} A_j A_r \\ \psi_0^{(n)} &= \sum_{ij} \alpha^{(n)}_{ij} A_i A_j + \sum_{ijr} \beta^{(n)}_{ijr} A_i A_i A_r \end{aligned}\right\} \quad (2.7.5)$$

The summations in Eqs. (2.7.5) are now defined with the convention $i \neq n$, $j \neq n$, $r \neq n$.

The expressions $\psi_0^{(n)}$, $\psi_1^{(n)}$ and $\psi_2^{(n)}$ represent the effects of the interactions of the amplitude A_n with amplitudes A_m, $m \neq n$.

Chapter 3

BEHAVIOUR OF SOLUTIONS

3.1 Formal simplification of the system of equations

Eqs. (2.7.3) describe how the amplitudes A_n develop as functions of time starting from some initial value. We shall particularly be interested in their behaviour in the limit as $t \to \infty$, and in the question whether in that limit the amplitude functions approach finite, stationary values. Inverting the question, we can first construct some stationary solutions of Eqs (2.7.3) and then investigate under what conditions these solutions represent the limit of A_n as $t \to \infty$. This is the procedure that we shall ultimately follow, but it should be pointed out at the outset that a complete exact solution of Eqs. (2.7.3), or even of the equations governing stationary solutions, must be considered beyond expectation. We shall hence be concerned with methods of approximation and in a later chapter we shall develop asymptotic methods of solution applicable in certain cases. However, for a better understanding of the phenomena involved some exploratory analysis is useful.

In the present chapter we shall attempt to learn about the behaviour of solutions as functions of time, by studying simplified systems of equations. If, in fact, we consider Eqs. (2.7.3) as a formal system, forgetting for a moment the procedure by which it was obtained, then we may take some liberty in chosing the coefficients of the system, and we may define these coefficients in such a way that solutions can be constructed without difficulty. If, in particular, we chose the coefficients so that $\psi_0^{(n)}$, $\psi_1^{(n)}$ and $\psi_2^{(n)}$ are only functions of A_m, $m < n$, then starting with the equation for $n = 0$ the simplified equations can be solved one by one.

Obviously, such a simplified system of equations represents a mathematical model which does not correspond to any partial non-linear differential equation of the class (2.2.1); nevertheless it will permit us to identify and to study certain interesting types of non-linear behaviour, which will reappear later on, in the analysis of the general system of Eqs. (2.7.3).

3.2 Stable and unstable stationary solutions

In the simplified system we have for A_0 the equation

$$\frac{dA_0}{dt} = \beta^{(0)} A_0^3 + \alpha^{(0)} A_0^2 - \mu_0 A_0 \tag{3.2.1}$$

From a somewhat more general point of view, Eq. (3.2.1) could be called the 'equation of an isolated amplitude', since it represents a problem in which interactions between different amplitudes have been removed. In a somewhat different form, an equation of this type was first introduced into the stability theory by L. D. LANDAU (1944), who conjectured that it should govern the amplitude of an unstable perturbation (see also LANDAU and LIFSCHITZ).

The stationary solutions of Eq. (3.2.1) are given by the roots of the right-hand-side, that is by

$$\overset{\infty}{A}{}_0^{(1,2)} = \frac{1}{2\beta^{(0)}}\{-\alpha^{(0)} \mp \sqrt{[\alpha^{(0)}]^2 + 4\mu_0 \beta^{(0)}}\}; \qquad \overset{\infty}{A}{}_0^{(3)} = 0 \tag{3.2.2}$$

We rewrite Eq. (3.2.1) in the form

$$\frac{dA_0}{dt} = \beta^{(0)} \left[A_0 - \overset{\infty}{A}{}_0^{(1)}\right] \left[A_0 - \overset{\infty}{A}{}_0^{(2)}\right] \left[A_0 - \overset{\infty}{A}{}_0^{(3)}\right] \tag{3.2.3}$$

and to analyse the behaviour of the solutions we utilize the phase-plane, that is, plot dA_0/dt as a function of A_0.

Evidently, the behaviour depends on the values of the coefficients μ_0, $\alpha^{(0)}$ and $\beta^{(0)}$. We shall suppose $\beta^{(0)} < 0$; at the end of this section we shall comment on the case $\beta^{(0)} > 0$.

We consider first the case $\mu_0 < 0$: in the linearized theory A_0 is an unstable perturbation which grows without limit as a function of time. The phase-plot for this case is sketched in fig. 1, where the arrows

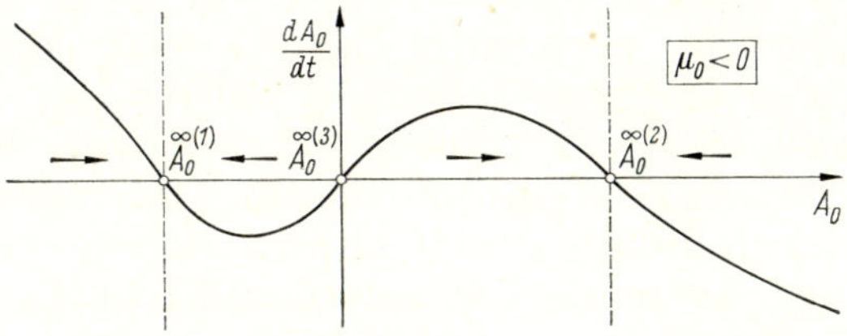

Figure 1

indicate the direction in which A_0 changes with time, in each interval of the A_0-axis.

It follows that:

$$\lim_{t\to\infty} A_0 = \overset{\infty}{A}{}_0^{(1)} \qquad \text{if for} \quad t=0 \qquad A_0 < 0$$

$$\lim_{t\to\infty} A_0 = \overset{\infty}{A}{}_0^{(2)} \qquad \text{if for} \quad t=0 \qquad A_0 > 0$$

$$\lim_{t\to\infty} A_0 = \overset{\infty}{A}{}_0^{(3)} \qquad \text{if for} \quad t=0 \qquad A_0 = 0$$

The stationary solution $\overset{\infty}{A}{}_0^{(3)} = 0$ is unstable, in agreement with the prediction of the linearized theory. The stationary solutions $\overset{\infty}{A}{}_0^{(1,2)}$ are both stable; from any arbitrary initial value the perturbation tends in the limit $t \to \infty$ to one of these stable stationary solutions. This behaviour is of course in sharp contrast to the behaviour of the linearized solution.

We can also integrate Eq. (3.2.1) directly; the simplest result is obtained when $\alpha^{(0)} = 0$, we then find:

$$A_0^2 = \frac{a_0^2\, e^{-2\mu_0 t}}{1 + \dfrac{a_0^2}{\overset{\infty}{A}{}_0^2}\left(e^{-2\mu_0 t} - 1\right)} \tag{3.2.5}$$

where a_0 is the initial value of A_0 and $\overset{\infty}{A}_0$ stands for $\overset{\infty}{A}{}_0^{(1)}$, or $\overset{\infty}{A}{}_0^{(2)}$. It is seen that at sufficiently small initial times, that is, as long as

$$\frac{a_0^2}{\overset{\infty}{A}{}_0^2}\left(e^{-2\mu_0 t} - 1\right) \ll 1$$

A_0 does behave in accordance with the linearized theory; however, in the limit $t \to \infty$ we have $A_0^2 \to \overset{\infty}{A}{}_0^2$.

Consider next the case $\mu_0 > 0$: in the linearized theory we have stability and the perturbation tends to zero as $t \to \infty$. If μ_0 is not too small, that is if

$$[\alpha^{(0)}]^2 < |4\mu_0 \beta^{(0)}|$$

then in Eq. (3.2.2) $\overset{\infty}{A}{}_0^{(1.2)}$ are complex and one easily establishes that in this case

$$\lim_{t\to\infty} A_0 = \overset{\infty}{A}{}_0^{(3)} = 0$$

for all initial values of A_0.

If, however, μ_0 is sufficiently small, that is, if

$$[\alpha^{(0)}]^2 > |4\mu_0 \beta^{(0)}|$$

then the three stationary solutions (3.2.2) are all real, and in the phase-plane we obtain the result sketched in fig. 2. (fig. 2 holds for $\alpha^{(0)} > 0$; for $\alpha^{(0)} < 0$ the direction of the A_0-axis should be reversed. Arrows again indicate the direction in which A_0 develops as a function of time).

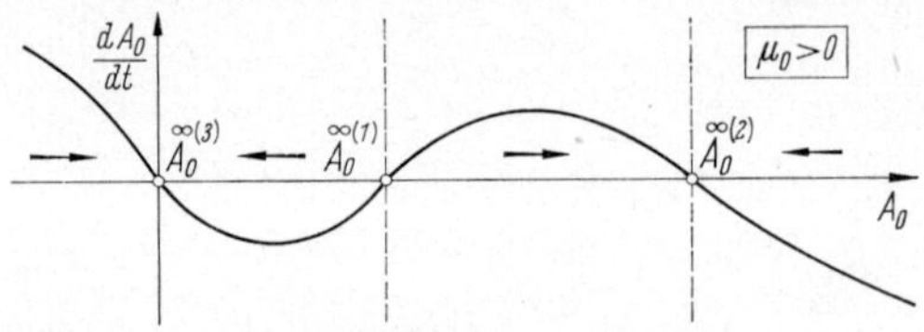

Figure 2

We see that:

$$\lim_{t\to\infty} A_0 = \overset{\infty}{A}{}_0^{(3)} = 0 \qquad \text{if for } t = 0 \quad A_0 < \overset{\infty}{A}{}_0^{(1)}$$

$$\lim_{t\to\infty} A_0 = \overset{\infty}{A}{}_0^{(2)} \qquad \text{if for } t = 0 \quad A_0 > \overset{\infty}{A}{}_0^{(1)}$$

The stationary solution $\overset{\infty}{A}{}_0^{(1)}$ is thus unstable. Moreover, it represents a critical value for the amplitude of perturbation: for an initial value smaller than this critical value perturbation tends to zero in the limit $t \to \infty$, for an initial value exceeding the critical value the perturbation tends in the limit $t \to \infty$ to a stationary, finite non-zero solution. In this example we have stability with respect to infinitesimal perturbations (in agreement with linear theory) but instability to perturbations of finite size.

More generally, in the case $\beta^{(0)} < 0$, we have found that there exists either one stable stationary solution, or two stable stationary solutions separated on the A_0-axis by an unstable one; the perturbations always remain bounded in the limit $t \to \infty$. In the case $\beta^{(0)} > 0$ one establishes by a similar analysis that the behaviour is reversed: there exist either one unstable solution, or two unstable solutions separated on the A_0-axis by a stable one; consequently, it is always possible to define initial values such that the perturbation grows without limit. This behaviour does not present much interest, in particular if one holds the opinion that a properly posed stability problem should not admit perturbations growing to infinite size. In what follows we shall usually assume $\beta^{(0)}$, and more generally $\beta^{(n)}$ to be negative.

3.3 Effects of interactions. Forced solutions

We now return to Eqs. (2.7.3), which we rewrite in a form analogous to (3.2.3):

$$\frac{dA_n}{dt} = \beta^{(n)} [A_n - A_n^{(1)}] [A_n - A_n^{(2)}] [A_n - A_n^{(3)}] \qquad (3.3.1)$$

where $A_n^{(1)}$, $A_n^{(2)}$ and $A_n^{(3)}$ are the three zeros of the polynomial of third degree in A_n, which forms the right-hand-side of Eq. (2.7.3). This polynomial contains the expressions $\psi_0^{(n)}$, $\psi_1^{(n)}$ and $\psi_2^{(n)}$, so that $A_n^{(1)}$, $A_n^{(2)}$ and $A_n^{(3)}$ are functions of A_m, $m \neq n$. However, in the simplified system of equations, $A_n^{(1)}$, $A_n^{(2)}$ and $A_n^{(3)}$ will only be functions of A_m, $m < n$. Thus, utilizing the results for A_0 of the preceeding section, we can in principle analyse completely the behaviour of A_1, with this result proceed to study A_2, etc. However, the analysis rapidly becomes very complicated, as the discussion below will show.

We remark that even in the simplified system, for $n > 0$, $A_n^{(1)}$, $A_n^{(2)}$ and $A_n^{(3)}$ are no longer constants, but functions of time. This makes the analysis in the phase-plane lose some of its attraction, since a new curve has to be traced at every instant of time. As $t \to \infty$, A_0 approaches one of its stable stationary solutions and $A_1^{(1)}$, $A_1^{(2)}$ and $A_1^{(3)}$ become the stationary solutions for A_1. Suppose that for A_0 we have the choice of two stable stationary solutions and suppose that for every one of them $A_1^{(1)}$, $A_1^{(2)}$ and $A_1^{(3)}$ are real; then for A_1 we would have six stationary solutions, some of which are stable and some unstable. As we proceed to A_2, A_3, etc. the number of possible stationary solutions increases, and in fact, for every A_n, we have three *classes* of stationary solutions $\overset{\infty}{A}{}_n^{(1)}$, $\overset{\infty}{A}{}_n^{(2)}$, $\overset{\infty}{A}{}_n^{(3)}$, every class containing a number of different solutions depending on the values of the amplitudes A_m, $m < n$, in the limit $t \to \infty$.

One class of solutions deserves particular interest and plays a fundamental role in most of the development which follows. From

Eqs. (3.3.1) and (2.7.3), by putting $dA_n/dt = 0$, we deduce

$$\overset{\infty}{A}{}_n^{(1)} \overset{\infty}{A}{}_n^{(2)} \overset{\infty}{A}{}_n^{(3)} = -\psi_0^{(n)} \frac{1}{\beta^{(n)}} \tag{3.3.2}$$

$$\overset{\infty}{A}{}_n^{(1)} + \overset{\infty}{A}{}_n^{(2)} + \overset{\infty}{A}{}_n^{(3)} = -[\alpha^{(n)} + \psi_2^{(n)}] \frac{1}{\beta^{(n)}} \tag{3.3.3}$$

$$\overset{\infty}{A}{}_n^{(1)} \overset{\infty}{A}{}_n^{(2)} + \overset{\infty}{A}{}_n^{(1)} \overset{\infty}{A}{}_n^{(3)} + \overset{\infty}{A}{}_n^{(2)} \overset{\infty}{A}{}_n^{(3)} = [-\mu_n + \psi_1^{(n)}] \frac{1}{\beta^{(n)}} \tag{3.3.4}$$

Clearly, one class of stationary solutions, —say $\overset{\infty}{A}{}_n^{(3)}$—, depends on $\psi_0^{(n)}$ in such way that $\overset{\infty}{A}{}_n^{(3)} \to 0$ if $\psi_0^{(n)} \to 0$. Now $\psi_0^{(n)}$ is a homogeneous function of A_m, $m \neq n$. It follows that $\overset{\infty}{A}{}_n^{(3)}$ has a functional dependence on A_m, $m \neq n$, with the property that $\overset{\infty}{A}{}_n^{(3)} \to 0$ as $A_m \to 0$ for all $m \neq n$. In other words, the solutions of the class $\overset{\infty}{A}{}_n^{(3)}$ take non-zero values only if *other* amplitudes A_m, $m \neq n$, take non-zero values. For obvious reasons, we shall call these solutions the 'forced solutions'. Note that in defining the forced solutions we did not make use of the simplified system of equations introduced in section 3.1; the existence of forced solutions is a property of the general system (2.7.3).

3.4 Analysis of forced solutions

As an illustration of the concept of forced solutions we analyse their appearance in the simplified system of equations. We consider the case $\mu_0 < 0$, $\mu_n > 0$ for $n > 0$. A_0 being determined from the equation of the isolated amplitude, we turn to the analysis of A_1.

The right-hand-side of Eq. (2.7.3) for $n = 1$ is a third degree polynomial in A_1, of which the coefficients are functions of A_0. $A_1^{(1)}$, $A_1^{(2)}$ and $A_1^{(3)}$ appear as roots of a cubic equation, whose general expressions are well-known. Since those formulas are not too pleasant to handle, we prefer to use series expansions. In fact, one easily shows that one root of the cubic is given by the power series

$$A_1^{(3)} = \sum_{k=0}^{\infty} \chi_k^{(1)} A_0^k \tag{3.4.1}$$

$$\chi_0^{(1)} = \chi_1^{(1)} = 0; \quad \chi_2^{(1)} = \frac{1}{\mu_1} \alpha_{00}^{(1)}; \quad \chi_3^{(1)} = \left[\beta_{000}^{(1)} + \frac{\alpha_{00}^{(1)} \alpha_{01}^{(1)}}{\mu_1}\right] \frac{1}{\mu_1}; \ \ldots \tag{3.4.2}$$

The series is convergent for sufficiently small values of A_0.

Let us assume presently that both initial value of A_0 and the stable stationary solutions $\overset{\infty}{A}{}_0^{(1,2)}$ are sufficiently small for the series (3.4.1) to converge. Then the series will be convergent at all times and $A_1^{(3)} \to \overset{\infty}{A}{}_1^{(3)}$

as $A_0 \to \overset{\infty}{A}{}_0^{(1\,2)}$. Eq. (3.4.1) thus represents in the limit $t \to \infty$ a series expansion for the forced solution of A_1.

We must still investigate under what conditions A_1 tends to the forced solution as $t \to \infty$. For that purpose $\overset{\infty}{A}{}_1^{(1.\,2)}$ have to be calculated. Now, $\overset{\infty}{A}{}_1^{(3)}$ being known from Eq. (3.4.1), $\overset{\infty}{A}{}_1^{(1.2)}$ appear as roots of a quadratic equation which can easily be analysed. One finds, as in the case of the isolated amplitude, that if μ_1 if not too small, $\overset{\infty}{A}{}_1^{(1.\,2)}$ are complex conjugates. In that case $\overset{\infty}{A}{}_1^{(3)}$ is the only stable stationary solution for A_1 and an analysis in the phase-plane shows that

$$\lim_{t\to\infty} A_1 = \overset{\infty}{A}{}_1^{(3)} \tag{3.4.3}$$

for any initial value of A_1.

But if μ_1 is small, then $\overset{\infty}{A}{}_1^{(1,2)}$ may be real and two stable stationary solutions for A_1 may exist, separated on the A_1-axis by an unstable solution. In the phase-plane that behaviour is indicated in fig. 3.

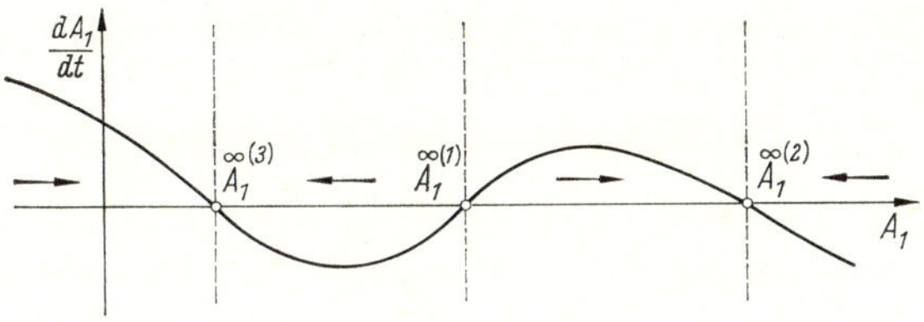

Figure 3

It follows that it is always possible to define initial values such that conclusion (3.4.3) remains true.

For certain well chosen initial values we hence have that $A_0 \to \overset{\infty}{A}{}_0^{(1,\,2)}$ and $A_1 \to \overset{\infty}{A}{}_1^{(3)}$ as $t \to \infty$. This being the case, we can proceed to study A_2. Again, the forced solution can be represented by a power series

$$\overset{\infty}{A}{}_2^{(3)} = \sum_{k=0}^{\infty} \chi_k^{(2)} \left[\overset{\infty}{A}{}_0^{(1,\,2)}\right]^k \tag{3.4.4}$$

$$\chi_0^{(2)} = \chi_1^{(2)} = 0; \qquad \chi_2^{(2)} = \frac{1}{\mu_2}\,\alpha_{00}^{(2)}; \;\ldots \tag{3.4.5}$$

By an analysis analogous to that for A_1 we can show that under certain conditions (certain initial values) we have

$$\lim_{t\to\infty} A_2 = \overset{\infty}{A}{}_2^{(3)} \tag{3.4.6}$$

Generalising the argument we can state that for every A_n, $n > 0$, the forced solution is represented by a power series

$$\overset{\infty}{A}{}_n^{(3)} = \sum_{k=0}^{\infty} \chi_k^{(n)} \left[\overset{\infty}{A}{}_0^{(1,2)}\right]^k \tag{3.4.7}$$

and $A_n \to \overset{\infty}{A}{}_n^{(3)}$ as $t \to \infty$, if the initial values of A_n satisfy certain conditions.

In the linearized theory, when $\mu_0 < 0$ while $\mu_n > 0$ for $n > 0$, A_0 grows without limit as a function of time and $A_n, n > 0$, all tend to zero. We see that non-linear effects, in the case described above, make A_0 tend to a finite value, while A_n, $n > 0$, all tend to forced solutions whose values are non-zero and are determined by the limiting value of A_0. We shall encounter this type of behaviour frequently in later developments.

3.5 Instability to finite size perturbations

In analyzing the equation of an isolated amplitude (section 3.2) we have already described a situation characterized by stability subject to infinitesimal perturbations and instability subject to disturbances of finite size. This phenomenon can further be complicated by the effects of interactions between different amplitudes A_n, as the example that follows will show. Let us consider a particularly simple system of equations

$$\frac{dA_n}{dt} + \mu_n A_n = \left\{\beta \sum_{m=0}^{n} A_m^2 + \sum_{m=0}^{n} \alpha_m^{(n)} A_m\right\} A_n \tag{3.5.1}$$

in which $\beta < 0$, $\alpha_m^{(n)} > 0$, $\alpha_m^{(0)} = 0$. We assume again $\mu_0 < 0$ and $\mu_n > 0$ for $n > 0$. The behaviour of A_0 is then defined by

$$\frac{dA_0}{dt} = \beta A_0\left(A_0^2 - \overset{\infty}{A}{}_0^2\right); \qquad \overset{\infty}{A}_0 = \mp\sqrt{\frac{\mu_0}{\beta}} \tag{3.5.2}$$

The behaviour of A_1 is described by

$$\frac{dA_1}{dt} = A_1 \beta (A_1 - A_1^{(1)}) (A_1 - A_1^{(2)}) \tag{3.5.3}$$

$$\begin{aligned} A_1^{(1)} &= \frac{-1}{2\beta}\left\{\alpha_1^{(1)} - \sqrt{[\alpha_1^{(1)}]^2 + 4\beta[\mu_1 - (\beta A_0^2 + \alpha_0^{(1)} A_0)]}\right\} \\ A_1^{(2)} &= \frac{-1}{2\beta}\left\{\alpha_1^{(1)} + \sqrt{[\alpha_1^{(1)}]^2 + 4\beta[\mu_1 - (\beta A_0^2 + \alpha_0^{(1)} A_0)]}\right\} \end{aligned} \tag{3.5.4}$$

It is seen that in the system of Eqs. (3.5.1) the forced solution for A_1 is non-existent, which greatly simplifies the analysis.

Now the values of $A_1^{(1,2)}$ depend on the values of the coefficients β, μ_1, $\alpha_0^{(1)}$, $\alpha_1^{(1)}$: $A_1^{(1,2)}$ may be real or complex, in the case of real values $A_0^{(1)}$ may be positive or negative. Thus, by a suitable choice of coefficients, various types of behaviour of the solution for A_1 may be realized. We shall restrict ourselves to the following, relatively simple case.

Let $\alpha_0^{(1)} = 2\sqrt{\mu_0 \beta}$ and let $[\alpha_1^{(1)}]^2 > 4|\beta|\mu_1$, $\mu_1 > |\mu_0|$. We assume the initial value of A_0 to be such that

$$0 \geqq A_0 \geqq \sqrt{\frac{\mu_0}{\beta}} \tag{3.5.5}$$

then from section 3.2 it follows that relation (3.5.5) holds at all times. Under those conditions $A_1^{(1,2)}$ are real, $A_1^{(1)} > 0$ and $A_1^{(2)} > A_1^{(1)}$. If we now consider $A_1^{(1)}$ as a function of A_0 we obtain the plot indicated in fig. 4.

In order to analyse the behaviour of A_1 we can again utilize the phase-plane. At any fixed instant of time we know the value of A_0 and hence the values of $A_1^{(1,2)}$. Plotting dA/dt as function of A_1 we obtain, at each instant of time, a result analogous to fig. 2. If now $A_1 > A_1^{(1)}$ then, at the instant under consideration, A_1 is growing, the behaviour is of an unstable type. If $A_1 < A_1^{(1)}$ then A_1 is decaying and the behaviour is of stable type. $A_1^{(1)}$ can hence be considered as an *instantaneous* critical value for instability.

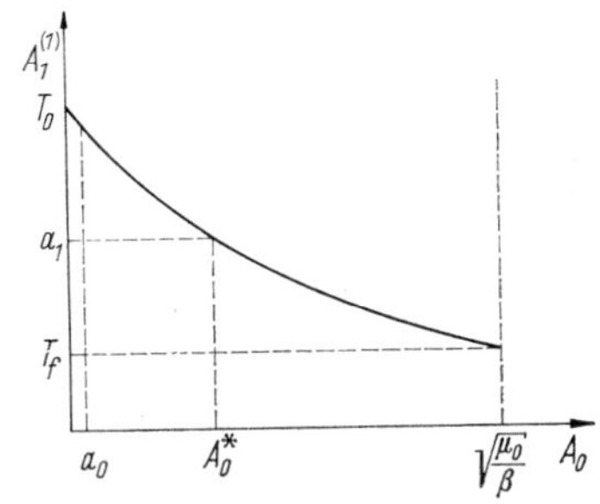

Figure 4

The phase-plot for A_1 will be different at every different instant of time and the behaviour of A_1 will depend on the initial values of perturbations and on the manner in which they are introduced. Consider now a stability problem in which perturbations are not only introduced at the initial instant, but may be reintroduced at any time $t > 0$. We then study the stability with respect to perturbations of a certain level, which may be introduced an arbitrary number of times during the time-history of the phenomenon. As we have stated in chapter 1, this formulation of a stability problem is in fact the only realistic one.

We define the levels $A_0 = a_0$, $A_1 = a_1$, which in the present example are chosen as indicated in fig. 4. Then A_0 starts by growing and A_1 decays. However, as A_0 develops toward its final value $\sqrt{\mu_0/\beta}$, the value of $A_1^{(1)}$ is being lowered and tends from T_0 to T_f. At some intermediate instant $t = t^*$, A_0 reaches the value A_0^* such that $[A_1^{(1)}] = a_1$. If now, at any instant $t > t^*$, a perturbation of the level $A_1 = a_1$ is reintroduced, then A_1 will grow instead of decaying and in the limit as $t \to \infty$ we shall have $A_0 \to \overset{\infty}{A_0}$ and $A_1 \to \overset{\infty}{A_1^{(2)}}$.

The behaviour of A_1, in the problem of stability with respect to perturbations of the level $A_1 = a_1$, is hence characterized by a transition from a stable solution to an unstable one which occurs when A_0 reaches a critical value.

3.6 Other types of behaviour

In the preceeding example, the interaction between A_0 and A_1 resulted in a lowering of the critical value for instability of A_1. Other more complicated types of behaviour, can be realized by different choise of coefficients of Eqs. (3.5.1). As an illustration let us briefly describe the following two cases.

a) We can have, as in section 3.5, $A_1^{(1,2)}$ real, $A_1^{(2)} > A_1^{(1)} > 0$, but $A_1^{(1)}$ such that as a function of A_0 the curve of A_1 displays a minimum in the interval $0 \leqq A_0 \leqq \overset{\infty}{A_0}$. With respect to perturbations of a level $A_1 = a_1$ the transition of solution for A_1 will then occur twice: as A_0 grows, first a transition from stable to unstable type occurs, followed, at higher values of A_0, by a transition from unstable to stable type.

b) We can have $A_1^{(2)} > A_1^{(1)} > 0$ in an interval of values of A_0, —say for $0 \leqq A_0 \leqq A_0'$—, while in the following interval, —say $A_0' \leqq A_0 \leqq A_0''$—, $A_1^{(2)} > 0$ but $A_1^{(1)} < 0$. In the interval $0 \leqq A_0 \leqq A_0'$ the behaviour of A_1 is characterized, as before, be the instantaneous critical value for instability $A_1^{(1)}$, while in the interval $A_0' \leqq A_0 \leqq A_0''$ we have instability to infinitesimal perturbations.

One can of course proceed to study A_2 in the presence of A_0 and A_1; A_3 in the presence of A_0, A_1, A_2; etc.; the complexity of behaviour and the richness in possibilities becoming ever greater. More generally, if one is free to choose the coefficients of Eq. (2.7.3) at will, then the number of cases that can be considered is only bound by the limits of one's imagination. However, as we have stated, we consider here simplified systems of equations which do not correspond to the non-linear stability problems defined in chapter 2. The study of these systems is only useful as a preliminary to the analysis of the general system of Eqs. (2.7.3), in which each A_n depends on *all* A_m, $m \neq n$. We have therefore limited the discussion of the present chapter to those examples of non-linear behaviour which illustrate the problems to be studied below.

Chapter 4

ASYMPTOTIC METHODS FOR PROBLEMS IN ONE-DIMENSIONAL SPACE

4.1 General outline

We now return to the system of Eqs. (2.7.2). The problem consists, first, of determining the stationary solutions of these equations. Then the nature of the stationary solutions must be analysed so that for given sets of initial conditions one may determine which stationary solution represents the limit of perturbations as $t \to \infty$. Most of this chapter will be concerned with the determination of stationary solutions by approximate methods. In a final section we shall analyse under what conditions the solutions that have been obtained represent the limit of perturbations for $t \to \infty$. In the present section we shall outline the concepts on which the developments of this chapter are based.

From Eqs. (2.7.2) it follows that stationary solutions $\overset{\infty}{A}_n$ must satisfy the system of equations

$$\sum_{ijr} \beta_{ijr}^{(n)} \overset{\infty}{A}_i \overset{\infty}{A}_j \overset{\infty}{A}_r + \sum_{ij} \alpha_{ij}^{(n)} \overset{\infty}{A}_i \overset{\infty}{A}_j - \mu_n \overset{\infty}{A}_n = 0; \quad n = 0, 1, 2, \ldots \quad (4.1.1)$$

A remark on terminology should be made here: we have called any stationary solution of the general problem formulated in section 2.2, a basic solution. Now studying the stability of a basic solution we are led to consider again stationary solutions for the perturbations. In this terminology, the basic solution under consideration plus any of the stationary perturbations gives another basic solution, and our analysis of behaviour is nothing but the study of the stability of this new basic solution.

It is clear that stationary solutions for perturbations can only exist if there is no uniqueness of the basic solution (see section 2.1).

It may thus seem rather arbitrary that we start from some basic solution and proceed by the stability theory to construct other basic solutions. In practice, however, we formulate the stability problem as described in section 2.2 by choosing for the basic solution some particularly simple stationary solution (one that can be found almost by inspection), while other stationary solutions if they exist, can not be obtained easily. From this point of view, the purpose of the analysis of this chapter is the construction of stationary solutions that are, in some sense, neighbouring the basic solution under consideration.

Note that we have left out of consideration the possibility of perturbations approaching in the limit $t \to \infty$ some non-stationary permanent-type solution. To study this eventuality under the most general conditions one has to admit complex coefficients in Eqs. (4.1.1), while the interpretation, in the context of one-dimensional problems, becomes somewhat more difficult. We shall however study such solutions later on in connection with two-dimensional problems (chapter 6 and sequel), where they arise in a natural way.

The solutions of Eqs. (4.1.1) depend on the values of the coefficients μ_n, $\alpha_{ij}^{(n)}$, $\beta_{ijr}^{(n)}$. In most cases, not all of these coefficients will be of the same order of magnitude, and some will be much smaller than others.

This suggests that we should search in Eqs. (4.1.1) for small parameters, so that solutions $\overset{\infty}{A}_n$ can be developed in asymptotic series with respect to these parameters. We shall introduce for this purpose, the formal transformation

$$\overset{\infty}{A}_n = \bar{\delta}_n a_n \tag{4.1.2}$$

We then obtain the system of equations

$$\sum_{ijr} \left[\frac{\beta_{ijr}^{(n)} \bar{\delta}_i \bar{\delta}_j \bar{\delta}_r}{\mu_n \bar{\delta}_n} \right] a_i a_j a_r + \sum_{ij} \left[\frac{\alpha_{ij}^{(n)} \bar{\delta}_i \bar{\delta}_j}{\mu_n \bar{\delta}_n} \right] a_i a_j - a_n = 0 \tag{4.1.3}$$

If we now determine the quantities $\bar{\delta}_n$ so that

$$\left| \frac{\beta_{ijr}^{(n)} \bar{\delta}_i \bar{\delta}_j \bar{\delta}_r}{\mu_n \bar{\delta}_n} \right| \leqq 1 ; \qquad \left| \frac{\alpha_{ij}^{(n)} \bar{\delta}_i \bar{\delta}_j}{\mu_n \bar{\delta}_n} \right| \leqq 1 \tag{4.1.4}$$

we obtain in Eqs. (4.1.3) a system of equations for the unknown quantities a_n, in which all coefficients are numerically at most of order unity. We can then reasonably expect that there will exist solutions a_n which will also be numerically of order unity. In other words, by the transformation (4.1.2) we attempt an a priori guess of the order of magnitude of the stationary solutions $\overset{\infty}{A}_n$.

In general it will not be possible to define $\bar{\delta}_n$ such that

$$\left| \frac{\beta_{ijr}^{(n)} \bar{\delta}_i \bar{\delta}_j \bar{\delta}_r}{\mu_n \bar{\delta}_n} \right| = 1, \qquad \left| \frac{\alpha_{ij}^{(n)} \bar{\delta}_i \bar{\delta}_j}{\mu_n \bar{\delta}_n} \right| = 1$$

for all values of n, i, j and r. Consequently, if the relations (4.1.4) are satisfied, we shall have in Eqs. (4.1.3) some coefficients equaling unity, while others are smaller. In fact, as we shall see shortly, Eq. (4.1.4) will in certain cases imply that while some coefficients equal unity, others are much smaller than unity. This opens the way to methods of approximation: asymptotic developments with respect to these small coefficients can be defined. Of course a posteriori one must verify that the solutions a_n obtained by this procedure are indeed of order unity.

In practical applications the values of the coefficients μ_n, $\alpha_{ij}^{(n)}$ and $\beta_{ijr}^{(n)}$ have to be known. In what follows we shall develop the method of analysis outlined above by considering certain classes of coefficients. In chosing these classes we attempt to cover cases of interesting behaviour, while preserving some simplicity. Examples of non-linear partial differential equations leading to stability problems with coefficients belonging to these classes will be presented and analysed in chapter 5.

We shall generally assume that $\beta_{ijr}^{(n)}$ are finite for all i, j, r and n. This being the case, the biggest of these coefficients can be chosen to normalize the others. We shall suppose for simplicity that $|\beta_{000}^{(0)}| \geqq |\beta_{ijr}^{(n)}|$ for all i, j, r and n, and we shall write

$$\bar{\beta}_{ijr}^{(n)} = \frac{\beta_{ijr}^{(n)}}{|\beta^{(0)}|}\,; \qquad \beta^{(0)} \equiv \beta_{000}^{(0)} \tag{4.1.5}$$

We then have

$$|\bar{\beta}_{ijr}^{(n)}| \leqq 1 \tag{4.1.6}$$

Of course, if a coefficient other then $\beta^{(0)}$ happens to be the largest $\beta_{ijr}^{(n)}$, an analogous normalisation can be performed.

We shall consider various classes of coefficients $\alpha_{ij}^{(n)}$, to be specified below; with respect to the eigenvalues μ_n we restrict ourselves to the case

$$\mu_n > 0 \quad \text{for} \quad n > 0 \tag{4.1.7}$$

Thus, cases of stability and instability are considered, but in case of instability the value of the parameter R, introduced in section 2,3, can not be too much in excess of R_{cr}.

It will also in general be necessary, for reasons of convergence, to assume that

$$\mu_n \to \infty \quad \text{as} \quad n \to \infty \tag{4.1.8}$$

in a sufficiently rapid way.

The asymptotic expansions that will be studied, will often be concerned with the case of small values of $|\mu_0|$. Now $|\mu_0| \to 0$ as $R \to R_{cr}$; hence these developments are valid in the vicinity of neutral stability. We shall speak of 'weak instability' when studying developments valid for $|\mu_0| \to 0$, $\mu_0 < 0$ (that is when $R \to R_{cr}$, $R - R_{cr} > 0$); we shall call 'weak stability' the case $|\mu_0| \to 0$, $\mu_0 > 0$ (and hence $R \to R_{cr}$, $R - R_{cr} < 0$). On the other hand, when $|\mu_0|$ is not particularly small, we shall denote by 'simple developed instability' the case $|\mu_0| \approx \mu_1$, $\mu_0 < 0$, $\mu_1 > 0$.

In the analysis that follows a somewhat modified form of transformation (4.1.2) proves to be more convenient. We shall generally write

$$\overset{\infty}{A}_n = \varepsilon\, \delta_n\, a_n \tag{4.1.9}$$

where ε is a parameter which measures the level of $\overset{\infty}{A}_n$. We can arrange this transformation in such way that

$$|\delta_n| \leqq 1 \tag{4.1.10}$$

In other words, ε is chosen of order of magnitude of the largest of parameters $\bar{\delta}_n$ used in the original transformation (4.1.2).

In the asymptotic expansions we shall use the usual 'order of magnitude' symbol, which should be read as follows:

$$f = 0(1) \quad \text{means} \quad \lim_{\varepsilon\to 0} [f] \quad \text{is finite}$$

$$f = 0(\varepsilon) \quad \text{means} \quad \lim_{\varepsilon\to 0} \left[\frac{1}{\varepsilon} f\right] \quad \text{is finite}$$

etc.

However, somewhat inconsistently, we shall also occasionally speak of *numerical* orders of magnitude, and by this we shall simply mean numerical values.

4.2 Weak stability or instability: the case $\alpha_{ij}^{(n)} = 0$

The simplest form of Eqs. (4.1.1) arises when $\alpha_{ij}^{(n)} = 0$. We then have

$$\overset{\infty}{A}_n = \frac{|\beta^{(0)}|}{\mu_n} \sum_{ijr} \bar{\beta}_{ijr}^{(n)} \overset{\infty}{A}_i \overset{\infty}{A}_j \overset{\infty}{A}_r \tag{4.2.1}$$

We seek a transformation (4.1.9) such that the system of equations for a_n will have coefficients not exceeding unity. One easily verifies that for this purpose one should choose

$$\overset{\infty}{A}_n = \varepsilon\, \delta_n\, a_n; \qquad \varepsilon = \sqrt{\left|\frac{\mu_0}{\beta^{(0)}}\right|}; \qquad \delta_n = \left|\frac{\mu_0}{\mu_n}\right| \tag{4.2.2}$$

The system of Eqs. (4.2.1) becomes then

$$a_n = \frac{\mu_n}{|\mu_n|} \sum_{ijr} \bar{\beta}_{ijr}^{(n)} \delta_i\, \delta_j\, \delta_r\, a_i\, a_j\, a_r \tag{4.2.3}$$

If we now consider weak stability or instability, characterized by $|\mu_0| \to 0$, then δ_n for $n > 0$ all become extremely small. We can therefore consider δ_n, $n > 0$, as small parameters and define asymptotic expansions with respect to these quantities. For this purpose we rewrite the sum occuring in Eq. (4.2.3) as follows:

$$\sum_{ijr} \bar{\beta}_{ijr}^{(n)} \delta_i\, \delta_j\, \delta_r\, a_i\, a_j\, a_r = \sum_{ijr}^{m-1} \bar{\beta}_{ijr}^{(n)} \delta_i\, \delta_j\, \delta_r\, a_i\, a_j\, a_r + R_m^{(n)} \tag{4.2.4}$$

Where the symbol $\sum\limits_{ijr}^{m-1}$ denotes a finite sum, defined by the inequalities $i \leqq m-1$; $j \leqq m-1$; $r \leqq m-1$.

The remainder-term $R_m^{(n)}$ is given by

$$\left.\begin{aligned} R_m^{(n)} = {} & \sum_{i=0}^{\infty} \sum_{j=0}^{\infty} \sum_{r=m}^{\infty} \bar{\beta}_{ijr}^{(n)} \delta_i\, \delta_r\, \delta_j\, a_i\, a_j\, a_r \\ & + \sum_{i=0}^{\infty} \sum_{j=m}^{\infty} \sum_{r=0}^{r=m-1} \bar{\beta}_{ijr}^{(n)} \delta_j\, \delta_j\, \delta_r\, a_i\, a_j\, a_r \\ & + \sum_{i=m}^{\infty} \sum_{j=0}^{j=m-1} \sum_{r=0}^{r=m-1} \bar{\beta}_{ijr}^{(n)} \delta_i\, \delta_j\, \delta_r\, a_i\, a_j\, a_r \end{aligned}\right\} \tag{4.2.5}$$

We consider the sum

$$S_m^{(n)} = \sum_{i=0}^{\infty} \sum_{j=0}^{\infty} \sum_{r=m}^{\infty} |\bar{\beta}_{ijr}^{(n)}| \, \delta_i \, \delta_j \frac{\delta_r}{\delta_m} \tag{4.2.6}$$

and assume that $S_m^{(n)}$ is finite for all values of n and m. For this it is necessary and sufficient that as $n \to \infty$, δ_n tends to zero (and hence μ_n tends to infinity) sufficiently rapidly. If, for example, $\delta_n \approx (n+1)^{-2}$, we would have $S_m^{(n)} \leqq (\pi^3/6)^3$. In general

$$S_m^{(n)} \leqq M \tag{4.2.7}$$

where M is a number independent of n and m. Under these conditions, and assuming always that the solutions a_n are finite, we can write for instance

$$\left| \sum_{i=0}^{\infty} \sum_{j=0}^{\infty} \sum_{r=m}^{\infty} \bar{\beta}_{ijr}^{(m)} \, \delta_i \, \delta_j \, \delta_r \, a_i \, a_j \, a_r \right| \leqq \delta_m \, \overline{M} \tag{4.2.8}$$

where $\overline{M}$ is a number independent of n and m. Finally, therefore

$$|R_m^{(n)}| \leqq \delta_m \, \overline{\overline{M}} \tag{4.2.9}$$

where $\overline{\overline{M}}$ is a number independent of n and m.

If we now formally consider δ_m as a small parameter, that is $\delta_m \to 0$, Eqs. (4.2.3) become

$$a_n = \frac{\mu_n}{|\mu_n|} \sum_{ijr}^{m-1} \bar{\beta}_{ijr}^{(n)} \, \delta_i \, \delta_j \, \delta_r \, a_i \, a_j \, a_r + 0(\delta_m) \tag{4.2.10}$$

In the case of weak stability or instability we have $|\mu_0| \to 0$ and hence $\varepsilon \to 0$ while $\delta_n = 0(\varepsilon^2)$ for all $n > 0$. The asymptotic solution now immediately follows from Eqs. (4.2.10)

$$a_n = \bar{\beta}_{000}^{(n)} \, a_0^3 + 0(\varepsilon^2) \quad \text{for} \quad n > 0 \tag{4.2.11}$$

$$a_0 = \frac{\mu_0}{|\mu_0|} \frac{\beta^{(0)}}{|\beta^{(0)}|} a_0^3 + 0(\varepsilon^2) \tag{4.2.12}$$

We see that the stationary solutions a_n for $n > 0$ are forced solutions, functions of a_0. The result for a_0 depends on the sign of $\beta^{(0)}$; we consider here for completeness both the cases $\beta^{(0)} < 0$ and $\beta^{(0)} > 0$.

a) If $\beta^{(0)} < 0$, there exist for weak instability ($\mu_0 < 0$) the stationary solutions

$$a_0^{(1,2)} = \mp 1 + 0(\varepsilon^2) \tag{4.2.13}$$

On the basis of the discussion of chapter 3, $a_0^{(1,2)}$ may be expected to be stable stationary solutions. For weak stability ($\mu_0 > 0$) there exist no real solutions of Eq. (4.2.12) and hence no stationary solutions of the order of magnitude presumed in the transformation (4.2.2).

b) For $\beta^{(0)} > 0$ there are, for weak instability, no solutions of the order magnitude defined in transformation (4.2.2), while for weak stability we have again the result (4.2.13). However, on the basis of chapter 3, these solutions must be expected to be unstable and represent threshold values for instability.

We remark finally that for the system of equations (4.2.1) the asymptotic solution for a_0 is identical with the solution of the equation for the isolated amplitude of section 3.2.

4.3 Weak stability or instability: the case $\alpha_{00}^{(0)} = 0$

We shall now consider a more general class of problems, in which not all $\alpha_{ij}^{(n)}$ are equal to zero. However, the behaviour of these coefficients with respect to i, j and n has to be specified. We shall assume that the ratio

$$\alpha_{ij}^{(n)} / \sqrt{\mu_n}$$

is uniformly bound for all i, j and $n > 0$. The coefficients can then be normalized and a simple but sufficiently representative case arises when the reduced coefficient

$$\bar{\alpha}_{ij}^{(n)} = \frac{\alpha_{ij}^{(n)}}{\sqrt{|\mu_n|\,|\beta^{(0)}|}} \tag{4.3.1}$$

is numerically at most of order unity for all i, j and $n > 0$. We shall assume this to be the case, although for the purpose of asymptotic expansions, valid in weak stability and instability, it is sufficient to require $\bar{\alpha}_{ij}^{(n)}$ to be finite.

We consider now the following transformation:

$$\overset{\infty}{A}_n = \varepsilon\,\delta_n\,a_n; \qquad \varepsilon = \sqrt{\left|\frac{\mu_0}{\beta^{(0)}}\right|}; \qquad \delta_n = \sqrt{\left|\frac{\mu_0}{\mu_n}\right|} \tag{4.3.2}$$

Eqs. (4.1.1) become:

$$\delta_n \sum_{ijr} \bar{\beta}_{ijr}^{(n)}\,\delta_i\,\delta_j\,\delta_r\,a_i\,a_j\,a_r + \sum_{ij} \bar{\alpha}_{ij}^{(n)}\,\delta_i\,\delta_j\,a_i\,a_j - \frac{\mu_n}{|\mu_n|}\,a_n = 0 \tag{4.3.3}$$

Clearly, for $n > 0$ the coefficients of Eqs. (4.3.3) are of order unity; for $n = 0$ this is only true if

$$\alpha_{00}^{(0)} = 0 \tag{4.3.4}$$

In the present section we assume condition (4.3.4) satisfied.

When $|\mu_0| \to 0$, we have $\varepsilon \to 0$ and $\delta_n = 0(\varepsilon)$ for $n > 0$. As in the preceding section, we assume that the sum $S_m^{(n)}$ exists; similarly we assume the sum

$$\sum_{i=0}^{\infty} \sum_{i=m}^{\infty} \bar{\alpha}_{ij}^{(n)}\,\delta_i\,\frac{\delta_j}{\delta_m}$$

to be finite for all n and m. Again, to satisfy these conditions we must require δ_n to approach zero sufficiently rapidly as $n \to \infty$.

The asymptotic solutions for $n > 0$ now follow immediately from Eqs (4.3.3) and we find

$$a_n = \bar{\alpha}_{00}^{(n)} a_0^2 + 0(\varepsilon); \quad n > 0 \tag{4.3.5}$$

These are obviously forced solutions. For $n = 0$ we take in Eq. (4.3.3) the limit $\varepsilon \to 0$, which leads to

$$\frac{\beta^{(0)}}{|\beta^{(0)}|} a_0^3 + \sum_{i=1}^{\infty} \frac{\alpha_{0i}^{(0)} + \alpha_{i0}^{(0)}}{\sqrt{|\mu_0||\beta^{(0)}|}} \sqrt{\left|\frac{\mu_0}{\mu_i}\right|} a_i a_0 - \frac{\mu_0}{|\mu_0|} a_0 = 0 \tag{4.3.6}$$

Now Eq. (4.3.6) contains not only a_0, but also all a_i, $i > 0$. However, we have just found that amplitudes other then a_0 are given by forced solutions and can be expressed in terms of a_0. Substituting for a_i, $i \neq 0$ the result (4.3.5) we obtain

$$\left[\frac{\beta^{(0)}}{|\beta^{(0)}|} + \gamma\right] a_0^3 - \frac{\mu_0}{|\mu_0|} a_0 = 0 \tag{4.3.7}$$

$$\gamma = \sum_{i=1}^{\infty} \frac{\alpha_{0i}^{(0)} + \alpha_{0i}^{(0)}}{\sqrt{|\mu_i||\beta^{(0)}|}} \frac{\alpha_{00}^{(i)}}{\sqrt{|\mu_i||\beta^{(0)}|}} \tag{4.3.8}$$

In analysing the solutions of Eq. (4.3.7) we must again distinguish the cases $\beta^{(0)} < 0$ and $\beta^{(0)} > 0$. Thus, if $\beta^{(0)} < 0$ stationary solutions exist in the case of weak instability if $\gamma < 1$. We then have

$$a_0^{(1,2)} = \mp \frac{1}{\sqrt{1-\gamma}} \tag{4.3.9}$$

If, for $\beta^{(0)} < 0$, we have $\gamma > 1$, than stationary solutions exist for weak stability and are given by

$$a_0^{(1,2)} = \mp \frac{1}{\sqrt{\gamma - 1}} \tag{4.3.10}$$

Now the solutions are indeed finite if $\gamma \neq 1$ as $\varepsilon \to 0$. Thus, for a given problem, we must have either $\gamma > 1$ or $\gamma < 1$. Consequently, stationary solutions will exist either in weak stability or in weak instability.

Similar analysis for the case $\beta^{(0)} > 0$, shows that stationary solutions exist in weak instability if $\gamma < -1$, they exist in weak stability if $\gamma > -1$.

We remark that in the present case, unlike the case considered in section 4.2, the asymptotic solution for a_0 is different from the solution of the equation of isolated amplitude. The interaction of a_0 with amplitudes other then a_0 plays an important role in establishing the value of the stationary solution.

4.4 Weak stability and instability: the case $\alpha_{00}^{(0)} \neq 0$

The development of the preceding section does not apply if $\alpha_{00}^{(0)} \neq 0$. We next consider that case, and at the same time study a somewhat different type of asymptotic expansion.

Until now we have considered asymptotic expansions based on the small parameter $|\mu_0| \to 0$. Suppose however that some other coefficient of Eqs. (4.1.1) takes very small or very large values. This can also lead to asymptotic expansions. We consider here as an example the case $|\beta^{(0)}| \gg 1$. For the other coefficients we shall for simplicity assume that

$$\bar{\alpha}_{ij}^{(n)} = \frac{\alpha_{ij}^{(n)}}{\alpha^{(n)}} \tag{4.4.1}$$

is numerically at most of order unity for all i, j and n, while $\alpha^{(0)}$ too is of this order of magnitude (we recall that the abbreviation $\alpha^{(n)}$ means $\alpha_{nn}^{(n)}$).

As in the preceding section, we consider small values of $|\mu_0|$. However, we now define these values by comparing them to the values of $\beta^{(0)}$; specifically we require that the order of magnitude of $|\mu_0|$ should not exceed the order of magnitude of $[\beta^{(0)}]^{-1}$.

In terms of the normalized coefficients $\bar{\alpha}_{ij}^{(n)}$ and $\bar{\beta}_{ijr}^{(n)}$, Eqs. (4.1.1) take the form

$$|\beta^{(0)}| \sum_{ijr} \bar{\beta}_{ijr}^{(n)} \overset{\infty}{A}_i \overset{\infty}{A}_j \overset{\infty}{A}_r + \alpha^{(n)} \sum_{ij} \bar{\alpha}_{ij}^{(n)} \overset{\infty}{A}_i \overset{\infty}{A}_j - \mu_n \overset{\infty}{A}_n = 0 \tag{4.4.2}$$

We introduce the transformation

$$\overset{\infty}{A}_n = \varepsilon\, \delta_n\, a_n; \qquad \varepsilon = \frac{1}{|\beta^{(0)}|}; \qquad \delta_0 = 1 \tag{4.4.3}$$

and obtain

$$\varepsilon \left\{ \sum_{ijr} \bar{\beta}_{ijr}^{(n)} \delta_i\, \delta_j\, \delta_r\, a_i\, a_j\, a_r + \alpha^{(n)} \sum_{ij} \bar{\alpha}_{ij}^{(n)} \delta_i\, \delta_j\, a_i\, a_j \right\} - \mu_n\, \delta_n\, a_n = 0 \tag{4.4.4}$$

Finally, we define δ_n for $n > 0$ as the larger of the two ratios ε/μ_n and $\varepsilon\,\alpha^{(n)}/\mu_n$. It is clear that in both cases $\delta_n = 0(\varepsilon)$ for $n > 0$.

In the asymptotic expansions of the preceding sections the process of $\varepsilon \to 0$ could be visualized as the approach of neutral stability. In the present case such an interpretation is no long possible: ε is a small quantity because some coefficients of the equations happen to be large. It is true that we have supposed $|\mu_0|$ to be not larger than ε, so that in the formal limit $\varepsilon \to 0$ we must have $|\mu_0| \to 0$. However, for every small but finite ε we can still let $|\mu_0| \to 0$ independently of ε.

In a manner similar to sections 4.2, 4.3, we obtain by asymptotic expansion of Eqs. (4.4.4) with respect to ε forced solutions for a_n, $n > 0$, which in the present case are of no particular interest. For $n = 0$, asymptotic expansion leads to the equation

$$\frac{\beta^{(0)}}{|\beta^{(0)}|} a_0^3 + \alpha^{(0)} a_0^2 - \mu_0\, |\beta^{(0)}|\, a_0 = 0(\varepsilon) \tag{4.4.5}$$

of which the solutions are

$$\left.\begin{aligned} a_0^{(1)} &= \tfrac{1}{2}\left\{\alpha^{(0)} - \sqrt{[\alpha^{(0)}]^2 + 4\beta^{(0)}\mu_0}\right\} + 0(\varepsilon) \\ a_0^{(2)} &= \tfrac{1}{2}\left\{\alpha^{(0)} + \sqrt{[\alpha^{(0)}]^2 + 4\beta^{(0)}\mu_0}\right\} + 0(\varepsilon) \\ a_0^{(3)} &= 0(\varepsilon) \end{aligned}\right\} \tag{4.4.6}$$

These are, of course, the solutions of the equation of isolated amplitude (section 3.2); they suggest an interesting interpretation here. We further restrict discussion to the case $\beta^{(0)} < 0$, $\alpha^{(0)} > 0$.

We recall that in section 2.3 we had introduced a parameter R, to characterize the basic solution whose stability was studied; for $R < R_{cr}$ there is stability while for $R > R_{cr}$ there is instability.

Suppose now that we increase R continuously from values $R - R_{cr} < 0$ to $R - R_{cr} > 0$, that is, we consider what happens in passing from weak stability to weak instability.

First, for $\mu_0 > \frac{1}{4}\frac{[\alpha^{(0)}]^2}{|\beta^0|}$, $a_0^{(1)}$ and $a_0^{(2)}$ will be complex, but as R increases μ_0 decreases and for $\mu_0 < \frac{1}{4}\frac{[\alpha^{(0)}]^2}{|\beta^{(0)}|}$, $a_0^{(1)}$ and $a_0^{(2)}$ become real and $0 < a_0^{(1)} < a_0^{(2)}$. We then have a situation analogous to fig. 2, with $a_0^{(1)}$ being the critical value for instability and $a_0^{(2)}$ the stable stationary solution. With a further increase of R, μ_0 decreases further and the critical value for instability $a_0^{(1)}$ tends to zero as $R \to R_{cr}$. As we now pass to weak instability, $R - R_{cr} > 0$, μ_0 becomes negative and $a_0^{(1)} < 0$, while $a_0^{(2)} > 0$. We arrive thus at a situation analogous to fig. 1, with $a_0^{(1)}$ and $a_0^{(2)}$ being stable stationary solutions.

To complete the investigation of the case $\alpha_{00}^{(0)} \neq 0$ we next reconsider the behaviour of the solution in weak stability and instability by applying, for every fixed $\beta^{(0)}$, the formal asymptotic expansion $|\mu_0| \to 0$, along the lines of sections 4.2 and 4.3

Some reflection shows that the appropriate transformation for the present case reads

$$\left.\begin{aligned} \overset{\infty}{A}_n &= \varepsilon\,\delta_n\,a_n; \qquad \varepsilon = |\mu_0|; \qquad \delta_0 = 1 \\ \delta_n &= \frac{|\mu_0|}{\sqrt{\mu_n}} \quad \text{for} \quad n > 0 \end{aligned}\right\} \tag{4.4.7}$$

As $|\mu_0| \to 0$ we have $\varepsilon \to 0$ and $\delta_n = 0(\varepsilon)$ for $n > 0$. From Eqs. (4.4.2) for $n > 0$ we find

$$\delta_n\,|\beta^{(0)}| \sum_{ijr} \bar{\beta}_{ijr}\,\delta_i\,\delta_j\,\delta_r\,a_i\,a_j\,a_r + \frac{\alpha^{(n)}}{\sqrt{\mu_n}} \sum_{ij} \bar{\alpha}_{ij}^{(n)}\,\delta_i\,\delta_j\,a_i\,a_j - \frac{\mu_n}{|\mu_n|}\,a_n = 0 \tag{4.4.8}$$

from which, by taking the limit $\varepsilon \to 0$, forced solutions for a_n, $n > 0$ can easily be deduced. On the other hand, Eq. (4.4.2) for $n = 0$ leads to

$$\varepsilon \sum_{ijr} \beta^{(0)}_{ijr}\, \delta_i\, \delta_j\, \delta_r\, a_i\, a_j\, a_r + \sum_{ij} \alpha^{(0)}_{ij}\, \delta_i\, \delta_j\, a_i\, a_j - \frac{\mu_0}{|\mu_0|}\, a_0 = 0 \qquad (4.4.9)$$

whose asymptotic approximation reads

$$\alpha^{(0)}\, a_0^2 - \frac{\mu_0}{|\mu_0|}\, a_0 + 0(\varepsilon) = 0 \qquad (4.4.10)$$

We obtain in this way one and only one asymptotic solution

$$a_0 = \frac{\mu_0}{|\mu_0|}\, \frac{1}{\alpha^{(0)}} + 0(\varepsilon) \qquad (4.4.11)$$

Comparing now with previous results—Eqs. (4.4.6), transformation (4.4.3)—we easily identify solution (4.4.11) as the first term of the development of $a_0^{(1)}$ for vanishing μ_0, which demonstrates the agreement between the two asymptotic expansions.

Summarizing, we have in the present case, where $|\beta^{(0)}| \gg 1$ and $|\mu_0| \to 0$, for a_0 two stationary solutions of different orders of magnitude. One solution, of which the nature changes from 'threshold value for instability' to 'stable stationary solution' as μ_0 passes through zero, is of the order of magnitude of $|\mu_0|$. The other solution, which is a stable stationary solution, both in weak instability and in weak stability is of the order ot magnitude of $[\beta^{(0)}]^{-1}$.

4.5 Method of approximation for the case of simple developed instability

In all cases considered so far, the asymptotic expansion led, in the first approximation, to very simple results. This is no longer so when we study developed instability, that is when $|\mu_0|$ is not very small. In fact, true asymptotic expansions of the type considered previously can no longer be constructed and we are reduced to methods of approximation, whose concepts belong however to the domain of the asymptotic calculus. We shall presently consider such a method of approximation, taking as example the class of coefficients of section 4.3. We shall assume that the reduced coefficients

$$\bar{\beta}^{(n)}_{ijr} = \frac{\beta^{(n)}_{ijr}}{|\beta^{(0)}|}\,; \qquad \bar{\alpha}^{(n)}_{ij} = \frac{\alpha^{(n)}_{ij}}{\sqrt{|\mu_n|\,|\beta^{(0)}|}}$$

are numerically at most of the order of magnitude of unity for all values of i, j, r and n. The transformation introduced in section 4.3 then still appears useful. We write

$$\overset{\infty}{A}_n = \sqrt{\left|\frac{\mu_0}{\beta^{(0)}}\right|}\, \delta_n\, a_n\,; \qquad \delta_n = \sqrt{\left|\frac{\mu_0}{\mu_n}\right|} \qquad (4.5.1)$$

and consider the system of equations

$$\delta_n \sum_{ijr} \bar{\beta}_{ijr}^{(n)} \delta_i \, \delta_j \, \delta_r \, a_i \, a_j \, a_r + \sum_{ij} \bar{\alpha}_{ij}^{(n)} \delta_i \, \delta_j \, a_i \, a_j - \frac{\mu_n}{|\mu_n|} a_n = 0 \tag{4.5.2}$$

The case that we study is characterized by $\mu_0 < 0$, $\mu_n > 0$ for $n > 0$. However, $|\mu_0|$ and μ_1 are numerically of the same order of magnitude. Thus, while δ_0 equals unity, δ_1 is of the order of magnitude of unity. Moreover, since $\mu_{n+1} > \mu_n$ we have $\delta_n < 1$ for $n > 0$. All coefficients of Eqs. (4.5.2) are hence numerically at most of order unity, which leads us to expect that the solutions a_n will also be at most of the order of magnitude of unity.

On the other hand, μ_n grows without limit as $n \to \infty$ and consequently

$$\delta_n \to 0 \quad \text{for} \quad n \to \infty \tag{4.5.3}$$

For sufficiently high values of n the quantities δ_n can hence be considered as the natural small parameters of the problem.

We assume that δ_n tends to zero with increasing n sufficiently rapidly so that the sum $S_m^{(n)}$ of Eq. (4.2.6) converges. In that case we can write, repeating the analysis of section 4.2:

$$\sum_{ijr} \bar{\beta}_{ijr}^{(n)} \delta_i \, \delta_j \, \delta_r \, a_i \, a_j \, a_r = \sum_{ijr}^{m-1} \bar{\beta}_{ijr}^{(n)} \delta_i \, \delta_j \, \delta_r \, a_i \, a_j \, a_r + R_m^{(n)} \tag{4.5.4}$$

$$|R_m^{(n)}| \leqq \delta_m M \tag{4.5.5}$$

where M is a number independent of n and m. In other words, by chosing δ_m sufficiently small we can approximate the infinite sum $\sum\limits_{ijr}$ by a finite sum $\sum\limits_{ijr}^{m-1}$ with any desired degree of accuracy. The condition for the validity of this statement is that a_n be finite, and for practical application, where the choice of a δ_m 'sufficiently small' has to be made, we shall require a_n to be numerically of the order of magnitude of unity.

A similar analysis applies to the second sum appearing in Eq. (4.5.2) so that finally we can write

$$\delta_n \sum_{ijr}^{m-1} \bar{\beta}_{ijr}^{(n)} \delta_i \, \delta_j \, \delta_r \, a_i \, a_j \, a_r + \sum_{ij}^{m-1} \bar{\alpha}_{ij}^{(n)} \delta_i \, \delta_j \, a_i \, a_j - \frac{\mu_n}{|\mu_n|} a_n = E_m^{(n)} \tag{4.5.6}$$

$$|E_m^{(n)}| \leqq \delta_m \overline{M} \tag{4.5.7}$$

where $\overline{M}$ is a number independent of n and m. We shall symbolize statement (4.5.7) by writing

$$E_m^{(n)} = 0(\delta_m) \tag{4.5.8}$$

Let us denote by ε a positive number, of which the order magnitude may be chosen as small as we wish. Then considering the sequence δ_n we can define, for every choice of ε, a value m such that $\delta_n < \varepsilon$ for

$n \geqq m$. This being accomplished, we obtain for $n \geqq m$ from Eqs. (4.5.6) the approximation

$$a_n = \frac{\mu_n}{|\mu_n|} \sum_{ij}^{m-1} \bar{\alpha}_{ij}^{(n)} \delta_i \delta_j a_i a_j + 0(\varepsilon) \tag{4.5.9}$$

In other words, for n sufficiently high, a_n can be represented by forced solutions (4.5.9), the error committed being of the order of magnitude of δ_m. For a_n, $n < m$, we still have the complete Eqs. (4.5.2), where for a_n, $n \geqq m$, forced solutions (4.5.9) can now be introduced. The problem is thus reduced to a finite system of m equations for m unknown quantities, the value of m depending on the value of ε that has been chosen. Naturally, the smaller the value of ε, the higher the value of m.

In applying the method exposed above considerable further simplification can be achieved by observing that not *all* terms of the order of magnitude of ε or smaller have yet been neglected. Clearly, if $\delta_n < \varepsilon$ for $n \geqq m$, then $\delta_i \delta_j < \varepsilon$ for certain combinations of i and j including yet some values $i \leqq m - 1$ and $j \leqq m - 1$. Thus, for example, if $\delta_n = (n + 1)^z$ with $z < 0$, then $\delta_i \delta_j < \delta_m$ if $(i + 1)(j + 1) \geqq m$. We shall therefore define now by the symbol

$$\sum_{ij}^{(m)}$$

the finite sum with respect to i and j, restricted to those values of i and j for which $\delta_i \delta_j > \delta_m$.

Similarly, we shall denote by the symbol

$$\sum_{ijr}^{(m/n)}$$

the finite sum with respect to i, j, and r, restricted to those values of i, j, r, for which

$$\delta_i \delta_j \delta_r > \frac{\delta_m}{\delta_n}$$

Neglecting thus all terms that are numerically of the order of magnitude of ε or smaller, we have for the forced solutions $n \geqq m$ the result

$$a_n = \frac{\mu_n}{|\mu_n|} \sum_{ij}^{(m)} \bar{\alpha}_{ij}^{(n)} \delta_i \delta_j a_i a_j + 0(\varepsilon) \tag{4.5.10}$$

while for $n < m$ we have

$$\delta_n \sum_{ijr}^{(m/n)} \bar{\beta}_{ijr}^{(n)} \delta_i \delta_j \delta_r a_i a_j a_r + \sum_{ij}^{(m)} \bar{\alpha}_{ij}^{(n)} \delta_i \delta_j a_i a_j - \frac{\mu_n}{|\mu_n|} a_n = 0 \tag{4.5.11}$$

These considerations not only reduce the number of terms in the equations, but also, in many cases, reduce the number of equations. Suppose, in fact, the value of m fixed by the choice of ε. Then as we consider

Eqs. (4.5.11) for $n < m$, the new definition of the sums may imply that for say $n > m^*$, $m^* < m$, the sums only contain a_n, $n \leqq m^*$. If this is the case, then already for $n > m^*$ we have forced solutions, and the final number of equations to be solved is reduced from m to m^*. In chapter 5 we shall see on an example that a reduction from 10 to 2 equations can arise in this way.

In final analysis, the method exposed above is only practicable if a not too high accuracy is acceptable (that is, if ε is not chosen too small), since otherwise one is confronted with finite but large systems of coupled algebraic equations of third degree. We shall however see in chapter 5 that with moderate accuracy interesting results can be obtained.

4.6 Behaviour of solutions as functions of time

Having constructed stationary solutions we must investigate under what conditions these solutions represent the limit of perturbations as time tends to infinity. Following the lines of chapter 3, we analyse, as an example, the case studied in section 4.3.

We consider first the equations for A_n, $n > 0$:

$$\frac{dA_n}{dt} = \beta^{(n)} [A_n - A_n^{(1)}] [A_n - A_n^{(2)}] [A_n - A_n^{(3)}] \tag{4.6.1}$$

In order to analyse the behaviour in the phase plane we need the values of $A_n^{(1,2,3)}$ at any instant of time. Suppose that at the initial instant $t = 0$ the perturbations satisfy the relations

$$A_n = \varepsilon\, \delta_n \bar{A}_n ; \qquad |\bar{A}_n| \leqq 1 \tag{4.6.2}$$

where

$$\varepsilon = \sqrt{\left|\frac{\mu_0}{\beta^{(0)}}\right|} ; \qquad \delta_n = \sqrt{\left|\frac{\mu_0}{\mu_n}\right|} \tag{4.6.3}$$

Then, following the methods of section 4.3 we can effectively calculate the three zeros $A_n^{(1,2,3)}$ and in the asymptotic approximation $\varepsilon \to 0$ we find

$$A_n^{(3)} = \varepsilon\, \delta_n\, \bar{\alpha}_{00}^{(n)} [\bar{A}_0^2 + 0(\varepsilon)] \tag{4.6.4}$$

$$A_n^{(1,2)} = -\tfrac{1}{2} \frac{\sqrt{|\mu_n|\,|\beta^{(0)}|}}{\beta^{(n)}} \left\{ \bar{\alpha}^{(n)} \pm \sqrt{[\bar{\alpha}^{(n)}]^2 - 4 \frac{\beta^{(n)}}{|\beta^{(0)}|}} \right\} [1 + 0(\varepsilon)] \tag{4.6.5}$$

We see that $A_n^{(3)}$ is smaller by an order of magnitude than $A_n^{(1,2)}$, and in the phase-plane we obtain the result indicated in fig. 5. (We have sketched the case of $A_n^{(1,2)}$ real.)

It is clear that at the initial instant A_n tends toward $A_n^{(3)}$. Moreover, A_n will tend toward $A_n^{(3)}$ at any instant $t > 0$, at which the order of magnitude relation (4.6.2) remains satisfied. But while $A_n^{(1,2)}$ are constant in the asymptotic approximation, $A_n^{(3)}$ depends on A_0 and is hence a

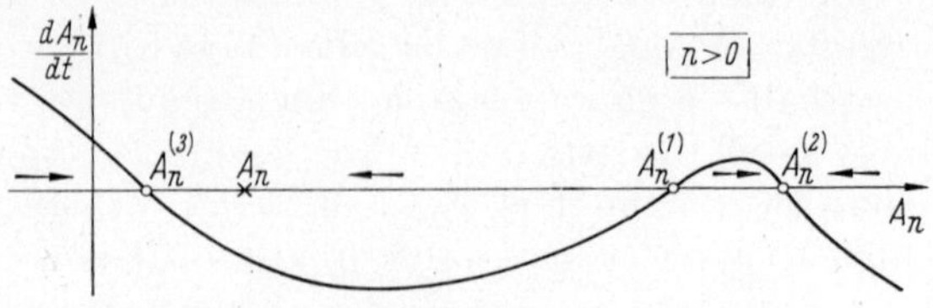

Figure 5

function of time. In fig. 5, the points $A_n^{(1,2)}$ are fixed, while points A_n and $A_n^{(3)}$ move as functions of time.

A more explicit result on the behaviour of A_n, $n > 0$, is obtained when introducing Eqs. (4.6.2), (4.6.4.), (4.6.5) into Eq. (4. 6.1). We then find

$$\frac{d\bar{A}_n}{dt} = \mu_n [\bar{A}_n - \bar{\alpha}_{00}^{(n)} \bar{A}_0^2] [1 + 0(\varepsilon)] \tag{4.6.6}$$

Integrating Eq. (4.6.6) we obtain after some simple algebra

$$\bar{A}_n = \bar{\alpha}_{00}^{(n)} \bar{A}_0^2 - e^{-\mu_n t} \left\{ \bar{\alpha}_{00}^{(n)} \int_0^t e^{\mu_n t'} \frac{d\bar{A}_0^2}{dt'} dt' - C_n \right\} \{1 + 0(\varepsilon)\} \tag{4.6.7}$$

where C_n are constants determined by the initial conditions.

Turning now to Eq. (4.6.1) for $n = 0$, one establishes without difficulty that at any instant at which condition (4.6.2) is satisfied one must have

$$\frac{d\bar{A}_0^2}{dt} = 0(\varepsilon^2) \tag{4.6.8}$$

This results simplifies Eq. (4.6.7) so that finally we obtain

$$\bar{A}_n = \bar{\alpha}_{00}^{(n)} \bar{A}_0^2 + C_n e^{-\mu_n t} + 0(\varepsilon) \tag{4.6.9}$$

Eq. (4.6.9) implies that although $A_n^{(3)}$ of Eq. (4.6.4) are not constants, in the asymptotic calculation they can be treated as such.

We can now substitute result (4.6.9) into Eq. (4.6.1) for $n = 0$ and analyse the behaviour of A_0. The solution is immediate and one finds

$$\lim_{t\to\infty} \bar{A}_0 = a_0^{(1,2)} \tag{4.6.10}$$

where $a_0^{(1,2)}$ are the stationary solutions determined in section 4.3. Consequently, for $n > 0$:

$$\lim_{t\to\infty} \bar{A}_n = \bar{\alpha}_{00}^{(n)} a_0^{(1,2)} \tag{4.6.11}$$

Thus, summarizing, we can state that for any initial values which do not surpass in order of magnitude the level defined in Eq. (4.6.2), (4.6.3), the perturbation tends in the limit $t \to \infty$ to the stationary solutions for which the asymptotic expansion has been given in section 4.3. It is seen that, among others, the infinitesimal perturbations belong to the class of admissible initial values. Moreover, the conclusion remains valid if in the course of time new perturbations are introduced, that is if for example in Eqs. (4.6.7) we change the constants C_n an arbitrary number of times and in an arbitrary fashion, provided that at all times the order of magnitude requirement (4.6.2), (4.6.3) is satisfied.

Chapter 5

ANALYSIS OF SOME ONE-DIMENSIONAL PROBLEMS

5.1 Introductory remarks

In the preceding chapters we have considered a rather wide class of non-linear partial differential equations and in developing methods of analysis for the non-linear stability problems we have made various assumptions, concerning the outcome of the linearized theory, the behaviour of coefficients, etc. We shall presently illustrate these developments by some specific examples, which show that the hypothesis which have been made are realistic, that is, that in certain problems of interest the assumptions are indeed verified.

On the other hand, since our results are concerned mostly with asymptotic expansions, a comparison with some exact solutions would be of fundamental interest. Unfortunately, this author was unable to construct a mathematical problem belonging to the class that has been defined, for which he could obtain an exact solution. We shall however compare the results, in one of the two problems treated below, to approximate solutions obtained by an entirely different method.

In present chapter we study first J. M. BURGERS' famous mathematical model of turbulence. This permits us to illustrate the developments of sections 4.3 and 4.5. In the final section we consider another simple mathematical model, which falls in the class of section 4.4.

5.2 Burgers' mathematical model of turbulence

In a series of papers, summarized in BURGERS (1948), the following problem was proposed and studied:

Let Φ be a function of η and t, which satisfies the equation

$$\frac{\partial \Phi}{\partial t} - \Phi - \frac{1}{R}\frac{\partial^2 \Phi}{\partial \eta^2} = -\frac{\partial \Phi^2}{\partial \eta} + U\Phi \tag{5.2.1}$$

where R is a constant, while U is a function of t, defined by

$$\frac{dU}{dt} + \frac{1}{R} U = -\int_0^1 \Phi^2 \, d\eta \tag{5.2.2}$$

The boundary conditions require that $\Phi = 0$ for $\eta = 0$ and $\eta = 1$.

In his analysis Burgers has demonstrated that the behaviour of Φ exhibits striking analogies with the behaviour of perturbations in hydrodynamic turbulence. In what follows we concentrate on the stability aspects of Burgers' model. First we summarize some of Burgers' results which will be needed in our analysis.

Let us seek stationary solutions of Eqs. (5.2.1), (5.2.2) subject to the given boundary conditions. Burgers has shown that for large values of the parameter R asymptotic solutions can be obtained; in the first approximation for $R \to \infty$ Burgers finds

$$\Phi \cong \tfrac{1}{2}(1 + U)\{\eta - \tanh[\tfrac{1}{2} R(1 + U)\,\eta]\} \tag{5.2.3}$$

$$\Phi \cong \tfrac{1}{2}(1 + U)\{(\eta - 1) - \tanh[\tfrac{1}{2} R(1 + U)(\eta - 1)]\} \tag{5.2.4}$$

These solutions are of 'boundary-layer type', and can further be analysed by decomposition in Fourier series.
We write:

$$\Phi(\eta) = \sum_{n=0}^{\infty} \overset{\infty}{A}_n \sin(n + 1)\,\pi\,\eta \tag{5.2.5}$$

Corresponding to solution (5.2.3) there is the spectrum

$$\overset{\infty}{A}_n \cong \frac{\pi}{R \sinh\left[\dfrac{\pi^2 (n + 1)}{R(1 + U)}\right]} \tag{5.2.6}$$

To the solution (5.2.4) corresponds the spectrum

$$\overset{\infty}{A}_n \cong \frac{(-1)^{n+1}\pi}{R \sinh\left[\dfrac{\pi^2 (n + 1)}{R(1 + U)}\right]} \tag{5.2.7}$$

Now Eqs. (5.2.3) to (5.2.6) still contain the quantity U, which has to be calculated from the formula

$$U = -R \int_0^1 \Phi^2 \, d\eta \tag{5.2.8}$$

It can be shown that for $R \to \infty$ one obtains

$$U \cong -1 + 2\sqrt{\frac{3}{R}} \tag{5.2.9}$$

5.3 Modification of Burgers' model. The problem of stability

We now modify Burgers model, by suppressing the term dU/dt in Eq. (5.2.2). Φ is then a solution of the equation

$$\frac{1}{R}\frac{\partial^2\Phi}{\partial\eta^2} + \Phi - \frac{\partial\Phi}{\partial t} = \frac{\partial\Phi^2}{\partial\eta} + R\left[\int_0^1 \Phi^2\, d\eta\right]\Phi \tag{5.3.1}$$

with boundary conditions

$$\Phi(0) = \Phi(1) = 0 \tag{5.3.2}$$

As result of this modification we have a problem belonging to the class defined in chapter 2. In the notation of that chapter the operator L is given by

$$L = \frac{1}{R}\frac{\partial^2}{\partial\eta^2} + 1 \tag{5.3.3}$$

the non-linear terms are given by

$$F(\Phi) = \left\{2\frac{\partial\Phi}{\partial\eta} + R\int_0^1 \Phi^2\, d\eta\right\}\Phi \tag{5.3.4}$$

On the other hand, it is clear that the stationary solutions of the modified and the original problem must be the same.

We formulate the stability problem by saying that we wish to investigate the stability of the trivial solution $\Phi = 0$. In Eq. (5.3.1) Φ can now be interpreted as the perturbation.

In the linearized analysis we neglect the right-hand-side of Eq. (5.3.1); to obtain the linearized solutions we write

$$\Phi = \sum_{n=0}^{\infty} b_n\, e^{-\mu_n t}\, \varphi_n(\eta) \tag{5.3.5}$$

we then find the eigenfunctions

$$\varphi_n = \sin(n+1)\,\pi\,\eta \tag{5.3.6}$$

and corresponding eigenvalues

$$\mu_n = -\left[1 - \frac{\pi^2}{R}(n+1)^2\right] \tag{5.3.7}$$

Clearly, the eigenvalues form an infinite denumerable sequence with the property $\mu_{n+1} > \mu_n$. Moreover, there exists a critical value $R_{cr} = \pi^2$ of the parameter R, such that for $R < R_{cr}$ there is stability while for $R > R_{cr}$ there is instability. Finally, for $\pi^2 \leq R \leq 4\pi^2$ we have $\mu_0 < 0$ and $\mu_n > 0$ for $n > 0$, which the case studied in chapter 4.

Turning to the non-linear problem we write the eigenfunction expansion

$$\Phi(\eta, t) = \sum_{n=0}^{\infty} A_n(t)\, \varphi_n(\eta) \tag{5.3.6}$$

Since L is a self-adjoint operator, the results of section 2.4 apply here. After some calculation we find

$$\frac{dA_n}{dt} + \mu_n A_n = -\frac{1}{2} R \left[\sum_{k=0}^{\infty} A_k^2\right] A_n +$$
$$+ (n+1)\,\pi \left[\frac{1}{2}\sum_{k=0}^{n-1} A_k A_{n-k-1} - \sum_{k=0}^{\infty} A_k A_{n+k+1}\right] \tag{5.3.7}$$

We remark that in the present case, due to the particularly simple form of the operator L, the eigenfunction expansion is identical with the Fourier series. It is from this last point of view that the system of equations (5.3.7) has been obtained by Burgers, who also discussed some of its properties (in particular those of the linearized solutions).

5.4 Asymptotic expansions in Burgers' model

The coefficients of Eqs. (5.3.7), as we can convince ourselves without difficulty, belong to the class defined in section 4.3, with $\alpha^{(0)} = 0$. Introducing the transformation

$$\overset{\infty}{A}_n = \varepsilon\,\delta_n\,a_n; \qquad \varepsilon = \sqrt{\frac{2\,|\mu_0|}{R}}; \qquad \delta_n = \sqrt{\left|\frac{\mu_0}{\mu_n}\right|} \tag{5.4.1}$$

we obtain the system of equations for the stationary solutions:

$$\frac{\mu_n}{|\mu_n|}\,a_n = -\,\delta_n^2 \left[\sum_{k=0}^{\infty} \delta_k^2\, a_k^2\right] a_n +$$
$$+ f_n \left[\tfrac{1}{2}\sum_{k=0}^{n-1} \delta_k\,\delta_{n-k-1}\,a_k\,a_{n-k-1} - \sum_{k=0}^{\infty} \delta_k\,\delta_{n+k+1}\,a_k\,a_{n+k+1}\right] \tag{5.4.2}$$

where

$$f_n = \frac{(n+1)\,\pi}{\sqrt{\frac{1}{2} R \left|1 - \frac{\pi^2}{R}(n+1)^2\right|}} \tag{5.4.3}$$

Comparing with section 4.3 we see that Eqs. (5.4.2) represent a simple case of the general system of equations (4.3.3). In fact, many of the coefficients $\alpha_{ij}^{(n)}$ and $\beta_{ijr}^{(n)}$ are equal to zero here, so that double and triple sums are reduced to simple sums.

It can be shown that sums $S_m^{(n)}$, etc. are finite, so that the asymptotic expansions outlined in chapter 4 are valid. For weak stability and instability we have

$$R \to \pi^2; \qquad \mu_0 \to 0; \qquad \varepsilon \to 0; \qquad \delta_n = 0(\varepsilon) \quad \text{for} \quad n > 0 \tag{5.4.4}$$

We obtain the forced solutions for a_n, $n > 0$, from Eqs. (5.4.2) by passing to limit $\varepsilon \to 0$:

$$a_1 = \sqrt{\tfrac{2}{3}}\, a_0^2 + 0(\varepsilon); \qquad a_n = 0(\varepsilon) \quad \text{for} \quad n > 1 \tag{5.4.5}$$

For $n = 0$ we find

$$\frac{\mu_0}{|\mu_0|} a_0 = -a_0^3 - f_0\,\delta_1\, a_0\, a_1 + O(\varepsilon) \tag{5.4.6}$$

Introduction of the forced solutions (5.4.5) leads to the equation

$$\frac{5}{3} a_0^3 + \frac{\mu_0}{|\mu_0|} a_0 + O(\varepsilon) = 0 \tag{5.4.7}$$

Solutions thus exist only for weak instability ($\mu_0 < 0$); in that case we find

$$\left.\begin{aligned} a_0 &= \mp \sqrt{\tfrac{3}{5}} + O(\varepsilon) \\ a_1 &= \tfrac{1}{5}\sqrt{6} + O(\varepsilon) \\ a_n &= O(\varepsilon) \quad \text{for} \quad n > 1 \end{aligned}\right\} \tag{5.4.8}$$

A somewhat unexpected feature of the present problem is that a_n for $n > 1$ come out to be of the order of magnitude of ε. This is due to the particular form of the coefficients $\bar{\alpha}_{ij}^{(n)}$ and $\bar{\beta}_{ijr}^{(n)}$; $\bar{\alpha}_{00}^{(n)}$, which yields the forced solutions in section 4.3, is identically zero here.

Modifying the procedure, slightly, we introduce a secondary transformation

$$a_n = \varepsilon^{n-1}\,\bar{a}_n \quad \text{for} \quad n \geqq 1 \tag{5.4.9}$$

We immediately find the forced solutions for a_n, $n > 1$, in the form of a recurrence formula

$$\bar{a}_n = \frac{1}{2} f_n \sqrt{\frac{2R}{\mu_{n-1}}}\,\bar{a}_{n-1}\, a_0 + O\,(\varepsilon) \tag{5.4.10}$$

A complete asymptotic solution for weak instability is thus obtained.

We now turn to the case of simple developed instability, that is the case of R-values such that $|\mu_0|$ and μ_1 are of the same order of magnitude. We shall follow in this the analysis of section 4.5 and perform the calculations taking as example the value $R = 2\pi^2$. We have in that case for δ_n the sequence

$$\delta_0 = 1; \quad \delta_1 = 0{,}706; \quad \delta_2 = 0{,}376; \quad \delta_3 = 0.288; \quad \text{etc.} \tag{5.4.11}$$

Thus, δ_n decays rather slowly with growing n, and if we chose $\varepsilon = 10^{-1}$ we have to go up to $m = 10$ in order to have $\delta_n < \varepsilon$ for $n \geqq m$. However, due to the particular form of the coefficients in Eqs. (5.4.2), and utilizing the developments described in the last part of section 4.5, we are able to obtain approximate solutions without excessive numerical labor.

We remark that for $\varepsilon = 10^{-1}$ not only $\delta_n < \varepsilon$ if $n \geqq 10$, but also $\delta_n\,\delta_m < \varepsilon$ if both $n \geqq 2$ and $m \geqq 2$. Neglecting thus in Eqs. (5.3.2) all terms that are numerically of the order of magnitude of ε or smaller, we obtain, as equivalent of Eqs. (4.5.11), for $n \geqq 2$:

$$a_n = f_n \left[\frac{1}{2}\sum_{k=0}^{n-1} \delta_k\,\delta_{n-k-1}\, a_k\, a_{n-k-1} - \sum_{k=0}^{k=1} \delta_k\,\delta_{n+k+1}\, a_k\, a_{n+k+1}\right] [1 + O(\varepsilon)] \tag{5.4.12}$$

Clearly, as in the case of weak instability, $a_n = 0(\delta_{n-1})$ for $n \geqq 2$. This allows us to neglect another group of terms in Eqs. (5.4.12), so that finally

$$a_n = \tfrac{1}{2} f_n \sum_{k=0}^{n-1} \delta_k\, \delta_{n-k-1}\, a_k\, a_{n-k-1} + 0(\varepsilon) \quad \text{for} \quad n \geqq 2 \quad (5.4.13)$$

For $a_n, n \geqq 2$, forced solutions are thus obtained. Turning to $n = 0$ and $n = 1$ now, we neglect in Eqs. (5.4.2) again all terms that are numerically of the order of magnitude of ε or smaller, and obtain

$$a_0\{-1 + a_0^2 + \delta_1^2 a_1^2 + 0(\varepsilon)\} = -f_0\{\delta_1 a_1 a_0 + \delta_1 \delta_2 a_1 a_2 + 0(\varepsilon)\} \quad (5.4.14)$$

$$a_1\{1 + \delta_1^2 a_0^2 + \delta_1^4 a_1^2 + 0(\varepsilon)\} = f_1\{\tfrac{1}{2} a_0^2 - \delta_2 a_0 a_2 + 0(\varepsilon)\} \quad (5.4.15)$$

Eqs. (5.4.14) and (5.4.15) still contain a_2, but the forced solution for this quantity is now available from Eq. (5.4.13), so that by substitution two coupled algebraic equations for a_0 and a_1 can be deduced. The solutions are:

$$a_0 = \mp 0.77; \qquad a_1 = 0.32 \qquad (5.4.16)$$

The forced solutions (5.4.13) can now be calculated; we find

$$a_2 = \pm 0.28; \qquad a_3 = 0.15; \qquad \text{etc.} \qquad (5.4.17)$$

It appears interesting to compare these numerical results with the values for $R = 2\pi^2$ which follow from Burgers' asymptotic formulas (section 5.2). Returning to the original notation for the amplitude functions—transformation (5.4.1)—we find:

$R = 2\pi^2$	$\pi \overset{\infty}{A}_0$	$\pi \overset{\infty}{A}_1$	$\pi \overset{\infty}{A}_2$	$\pi \overset{\infty}{A}_3$
Burgers' asymptotic result for $R \to \infty$.	$\mp$ 0.58	0.22	$\mp$ 0.097	0.045
Approximate solution with $\varepsilon = 10^{-1}$.	$\mp$ 0.55	0.16	$\mp$ 0.075	0.031

It is seen that the two results agree in numerical order of magnitude and in behaviour with n. Certainly, the approximate solution with ε chosen as 10^{-1} can not claim great accuracy. On the other hand, we have applied Burgers' asymptotic solution for $R \to \infty$ at the value $R \cong 20$. The agreement between the results, obtained by two methods of which the basic assumptions are entirely different, appears encouraging.

Returning now the case of weak instability, we remark that our asymptotic solution yields $\overset{\infty}{A}_n \to 0$ as $R \to \pi^2$; Burgers' development for $R \to \infty$, if applied formally at $R = \pi^2$. leads to values $\overset{\infty}{A}_n \neq 0$.

Thus, Burgers' asymptotic solution for $R \to \infty$, although still applicable at $R = 2\pi^2$, breaks down in the vicinity of $R = \pi^2$.

5.5 Another simple mathematical model

In a final example we consider the equation

$$\frac{1}{R}\frac{\partial^2 \Phi}{\partial \eta^2} + \frac{2}{\sqrt{R}}\frac{\partial}{\partial \eta}(\eta\,\Phi) + \eta^2\,\Phi - \frac{\partial \Phi}{\partial t} = F(\Phi) \tag{5.5.1}$$

with

$$F(\Phi) = \left[\frac{\partial \Phi}{\partial \eta} + R^2 \int_0^1 \Phi^2\,d\eta\right]\Phi \tag{5.5.2}$$

and with boundary conditions

$$\Phi = 0 \quad \text{for} \quad \eta = 0 \quad \text{and} \quad \eta = 1 \tag{5.5.3}$$

We again study the stability of the trivial solution $\Phi = 0$.

The present problem is in many ways analogous to Burgers' mathematical model of turbulence; however, the operator L is not self-adjoint. Moreover, we have the opportunity to illustrate certain phenomena which did not appear in Burgers' problem.

The linearized solutions of the stability problem are easily obtained; we find:

$$\Phi = \sum_{n=0}^{\infty} b_n\,e^{-\mu_n t}\,\varphi_n(\eta) \tag{5.5.4}$$

with eigenfunctions

$$\varphi_n = e^{-\frac{1}{2}\sqrt{R}\,\eta^2}\sin(n+1)\,\pi\,\eta \tag{5.5.5}$$

and eigenvalues

$$\mu_n = \frac{1}{R}\left[(n+1)^2\,\pi^2 - \sqrt{R}\right] \tag{5.5.6}$$

Again, μ_n is an infinite denumerable sequence and there exists a critical value $R_{cr} = \pi^4$ of the parameter R such that for $R < R_{cr}$ there is stability while for $R > R_{cr}$ there is instability. We shall be concerned, as before, with weak stability and instability.

Since the operator L is not self-adjoint, the eigenfunctions φ_n are not mutually orthogonal. However, by a simple transformation the operator can be brought into a self-adjoint form, which is illustrated by the fact that functions $\varphi_n \exp\left(\frac{1}{2}\sqrt{R}\,\eta^2\right)$ are orthogonal.

On the other hand, we can simply utilize the formalism of section 2.5 and define adjoint eigenfunctions

$$\tilde{\varphi}_n = 2e^{\frac{1}{2}\sqrt{R}\,\eta^2}\sin(n+1)\,\pi\,\eta \tag{5.5.7}$$

The functions φ_n and $\tilde{\varphi}_n$ are bi-orthogonal. Following section 2.5, we thus arrive at the system of equations

$$\frac{dA_n}{dt} + \mu_n A_n = \sum_{ij} \alpha_{ij}^{(n)} A_i A_j + \Big[\sum_{ij} \beta_{ij} A_i A_j\Big] A_n \qquad (5.5.8)$$

of which the coefficients $\alpha_{ij}^{(n)}$ and β_{ij} must be calculated from

$$\sum_{ij} \beta_{ij} A_i A_j = -R^2 \int_0^1 \Phi^2 \, d\eta \qquad (5.5.9)$$

$$\sum_{ij} \alpha_{ij}^{(n)} A_i A_j = -\frac{1}{2} \int_0^1 \frac{\partial \Phi^2}{\partial \eta} \tilde{\varphi}_n \, d\eta \qquad (5.5.10)$$

It is seen that Eqs. (5.5.8) are still simpler than the general system studied in chapter 4, no triple sums occur here and the coefficients β_{ij} do not depend upon n.

In the effective calculation of the coefficients no closed-form results can be obtained, unless one accepts the approximation

$$\int_0^1 e^{-\sqrt{R}\,\eta^2} \cos m\pi\eta \, d\eta \cong \frac{\sqrt{\pi}}{2R^{1/4}} e^{-\frac{m^2\pi^2}{4\sqrt{R}}} \left\{1 + 0\left(e^{-\sqrt{R}}\right)\right\} \qquad (5.5.11)$$

For the values of R of the order of magnitude of π^4 this approximation appears permissible. Performing the calculation we then obtain the following expressions:

$$\beta_{ij} = -R^2 \left[\frac{\sqrt{\pi}}{4R^{1/4}}\right] \bar{\bar{\beta}}_{ij}; \qquad \alpha_{ij}^{(n)} = (n+1)\,\pi \left[\frac{\sqrt{2\pi}}{4R^{1/4}}\right] \bar{\bar{\alpha}}_{ij}^{(n)} \qquad (5.5.12)$$

$$\bar{\bar{\beta}}_{ij} = \exp\left[-\frac{(i-j)^2\pi^2}{4\sqrt{R}}\right] - \exp\left[-\frac{(i+j+2)^2\pi^2}{4\sqrt{R}}\right] \qquad (5.5.13)$$

$$\begin{aligned}\bar{\bar{\alpha}}_{ij}^{(n)} = {} & \frac{n+i+2}{n+1}\left\{\exp\left[-\frac{(n+i-j+1)^2\pi^2}{2\sqrt{R}}\right] - \exp\left[-\frac{(n+i+j+3)^2\pi^2}{2\sqrt{R}}\right]\right\} \\ & + \frac{n-i}{n+1}\left\{\exp\left[-\frac{(n-i+j+1)^2\pi^2}{2\sqrt{R}}\right] - \exp\left[-\frac{(n-i-j+1)^2\pi^2}{2\sqrt{R}}\right]\right\}\end{aligned} \qquad (5.5.14)$$

Analysing the coefficients we see that if R is of the order of magnitude of π^4, β_{00} is numerically large (of the order 10^3), while the reduced coefficients

$$\bar{\alpha}_{ij}^{(n)} = \frac{\alpha_{ij}^{(n)}}{\alpha_{nn}^{(n)}}$$

are of the order of magnitude of unity, and so is the coefficient $\alpha_{00}^{(0)}$. We have thus the case studied in section 4.4, and we can consider asymptotic expansions with respect to the inverse of β_{00}. We introduce for this purpose the transformation

$$\overset{\infty}{A}_n = \varepsilon\, \delta_n\, a_n; \qquad \varepsilon = \frac{1}{|\beta_{00}|}; \qquad \delta_0 = 1 \tag{5.5.15}$$

Eqs. (5.5.8) for stationary solutions take the form

$$\varepsilon \left\{\delta_n \left[\sum_{ij} \frac{\bar{\bar{\beta}}_{ij}}{|\bar{\bar{\beta}}_{00}|}\, \delta_i\, \delta_j\, a_i\, a_j\right] a_n + \alpha_{nn}^{(n)} \sum \bar{\alpha}_{ij}^{(n)}\, \delta_i\, \delta_j\, a_i\, a_j\right\} - \mu_n\, \delta_n\, a_n = 0 \tag{5.5.16}$$

We chose for $n \geqq 1$

$$\delta_n = \varepsilon \frac{\alpha_{nn}^{(n)}}{\mu_n} \tag{5.5.17}$$

Evidently, $\delta_n = 0(\varepsilon)$ for $n \geqq 1$. Asymptotic expansion of Eqs. (5.5.16) with respect to ε yields forced solutions for $n \geqq 1$; for $n = 0$ we obtain the equation

$$-a_0^3 + \alpha_{00}^{(0)}\, a_0^2 - \mu_0\, |\beta_{00}|\, a_0 = 0 \tag{5.5.18}$$

of which the solutions are

$$\left.\begin{aligned} a_0^{(1)} &= \tfrac{1}{2}\left\{\alpha_{00}^{(0)} - \sqrt{[\alpha_{00}^{(0)}]^2 - 4\mu_0\, |\beta_{00}|}\right\} \\ a_0^{(2)} &= \tfrac{1}{2}\left\{\alpha_{00}^{(0)} + \sqrt{[\alpha_{00}^{(0)}]^2 - 4\mu_0\, |\beta_{00}|}\right\} \\ a_0^{(3)} &= 0 \end{aligned}\right\} \tag{5.5.19}$$

As in section 4.4, we now have the following interpretation:

Let R approach the value R_{cr} in weak stability. Then, for sufficiently small values of μ_0, the phenomen of instability to finite size perturbations appears, the critical amplitude of perturbations being given by $a_0^{(1)}$. As $R \to R_{cr}$, the critical value $a_0^{(1)} \to 0$. Finally, in weak instability, $a_0^{(1,2)}$ are the two stable stationary solutions

It is of some interest to evaluate the value of R for which instability to finite size perturbations makes it appearance. From Eq. (5.5.19) we see that $a_0^{(1,2)}$ become real when $|\mu_0| = \mu_0^*$, μ_0^* being given by

$$\mu_0^* = \frac{1}{4} \frac{[\alpha_{00}^{(0)}]^2}{|\beta_{00}|} \tag{5.5.20}$$

Utilizing the explicit formulas for μ_0, $\alpha_{00}^{(0)}$ and β_{00}, we find that $|\mu_0| = \mu_0^*$ for $R = R_{cr}^*$,

$$R_{cr}^* = R_{cr}\{1 - 0.16 R_{cr}^{-1}[1 + 0(R_{cr}^{-1})]\}; \qquad R_{cr} = \pi^4 \tag{5.5.21}$$

Summarizing, the behaviour can be described as follows:

For $R < R_{cr}^*$ there is stability with respect to perturbations whose level is limited by

$$\left.\begin{aligned} &A_n = \varepsilon\, \delta_n\, \bar{A}_n; \qquad |\bar{A}_n| \leqq 1 \\ &\varepsilon = \frac{1}{|\beta_{00}|}; \qquad \delta_0 = 1; \qquad \delta_n = 0(\varepsilon) \end{aligned}\right\} \tag{5.5.22}$$

For $R^*_{cr} < R < R_{cr}$, and with respect to perturbations of the level (5.5.22), instability arises if the amplitude exceeds a critical value.

For $R > R_{cr}$, finally, there is instability with respect to infinitesimal perturbations. In all cases of instability, however, the perturbation approaches a finite stationary value in the limit $t \to \infty$.

Chapter 6

A CLASS OF PROBLEMS IN TWO-DIMENSIONAL SPACE

6.1 Introductory remarks

We now turn to problems in which the unknown quantities are functions of two space variables and time. Naturally theses problems are of much greater complexity than the one-dimensional stability problems studied in the preceding chapters. We shall find however, at various points of our study, that many concepts introduced in the investigation of the one-dimensional problems still retain their utility. Indeed, from the point of view of mathematical analysis, the preceding chapters can almost be considered as an introduction to the study of problems in two-dimensional space.

The class of problems defined and studied here is in many respects a generalization of the class described in chapter 2, but in contrast to the one-dimensional case, some well-known physical stability problems are included in the class to be considered. They are, the hydrodynamic stability problems of POISEUILLE and COUETTE flow between parallel walls. We shall find at a later stage, that the restriction to the study of weak stability and instability excludes the case of COUETTE flow from further consideration. The study of the POISEUILLE flow problem will be presented in some detail in chapter 9, as an application of the general theory to be developed in the chapters that follow.

6.2 Formulation

Let ξ and η be the space-coordinates and t, as before, the time. We consider the function $\Phi(\xi, \eta, t)$, which is a solution of the equation

$$L(\Phi) - \frac{\partial}{\partial t} S(\Phi) = \sum_{l=1}^{N} P^{(l)}(\Phi)\, Q^{(l)}(\Phi) \qquad (6.2.1)$$

in the space domain $0 \leqq \eta \leqq 1$, $-\infty \leqq \xi \leqq +\infty$.

L, S, $P^{(l)}$ and $Q^{(l)}$ symbolize real linear operators, independent of time and invariant with respect to translation along the ξ-axis. The number N of bi-linear terms in Eq. (6.2.1) is supposed finite.

Clearly, Eq. (6.2.1) represents a generalization of Eq. (2.2.1) of the one-dimensional theory, but the form of non-linear terms is somewhat more restricted here, only bi-linear terms being admitted.

At $\eta = 0$ and $\eta = 1$ boundary conditions are imposed, which we shall assume to be linear, independent of time and invariant with respect to ξ. Finally, we shall require Φ to be real.

We assume that the problem admits at least one stationary solution $\Phi = \varphi_0(\eta)$, independent of the variable ξ. Such a solution satisfies the equation

$$L[\varphi_0(\eta)] = \sum_{l=1}^{N} P^{(l)}[\varphi_0(\eta)]\, Q^{(l)}[\varphi_0(\eta)] \tag{6.2.2}$$

and the boundary conditions of the problem. φ_0 is the basic solution of which the stability we wish to investigate.

As an illustration let us recall briefly the formulation of the problem of two-dimensional flow of an incompressible fluid between two parallel walls. We denote by u and v the fluid velocities in the ξ and η directions. The stream function Φ, with the properties

$$\frac{\partial \Phi}{\partial \eta} = u; \qquad \frac{\partial \Phi}{\partial \xi} = -v \tag{6.2.3}$$

is then a solution of the equation

$$\frac{1}{R} \nabla^2(\nabla^2 \Phi) - \frac{\partial}{\partial t} \nabla^2 \Phi = \frac{\partial \Phi}{\partial \eta} \frac{\partial \nabla^2 \Phi}{\partial \xi} - \frac{\partial \Phi}{\partial \xi} \frac{\partial \nabla^2 \Phi}{\partial \eta} \tag{6.2.4}$$

where R is the Reynolds-number, and where

$$\nabla^2 = \frac{\partial^2}{\partial \eta^2} + \frac{\partial^2}{\partial \xi^2} \tag{6.2.5}$$

Clearly, Eq. (6.2.4) belongs to the class defined above and in the notation of Eq. (6.2.1) can be written as

$$S = \nabla^2; \quad L = \frac{1}{R} S^2; \quad N = 2$$

$$P^{(1)} = -\frac{\partial}{\partial \xi}; \quad P^{(2)} = \frac{\partial}{\partial \eta}; \quad Q^{(1)} = \frac{\partial S}{\partial \eta}; \quad Q^{(2)} = \frac{\partial S}{\partial \xi} \tag{6.2.6}$$

In the case of Couette-flow, that is when the two walls move with respect to each other, the conditions at the walls may be of the form

$$\begin{aligned} \left(\frac{\partial \Phi}{\partial \eta}\right)_{\eta=0} &= 0; \qquad \left(\frac{\partial \Phi}{\partial \eta}\right)_{\eta=1} = 1 \\ \left(\frac{\partial \Phi}{\partial \xi}\right)_{\eta=0} &= 0; \qquad \left(\frac{\partial \Phi}{\partial \xi}\right)_{\eta=1} = 0 \end{aligned} \tag{6.2.7}$$

In the case of stationary walls (Poiseuille-flow) the boundary conditions are

$$\begin{aligned} \left(\frac{\partial \Phi}{\partial \eta}\right)_{\eta=0} &= \left(\frac{\partial \Phi}{\partial \eta}\right)_{\eta=1} = 0 \\ \left(\frac{\partial \Phi}{\partial \xi}\right)_{\eta=0} &= \left(\frac{\partial \Phi}{\partial \xi}\right)_{\eta=1} = 0 \end{aligned} \tag{6.2.8}$$

Basic solutions are easily found in both cases; for Couette-flow we have

$$\varphi_0 = \tfrac{1}{2}\,\eta^2 + \text{const.} \tag{6.2.9}$$

while for Poiseuille-flow we find

$$\varphi_0 = \tfrac{1}{2}\,\eta^2 - \tfrac{1}{3}\,\eta^3 + \text{const.} \tag{6.2.10}$$

6.3 The problem of stability. Linearized theory

To investigate the stability of a basic solution φ_0 of an equation of the class (6.2.1) we write

$$\Phi(\xi, \eta, t) = \varphi_0(\eta) + \Phi'(\xi, \eta, t) \tag{6.3.1}$$

where Φ' is the perturbation. By substitution, and using Eq. (6.2.2), we find that Φ' must satisfy the equation

$$L(\Phi') - \frac{\partial}{\partial t} S(\Phi') - \sum_{l=1}^{N} [P^{(l)}(\varphi_0)\, Q^{(l)}(\Phi') + Q^{(l)}(\varphi_0)\, P^{(l)}(\Phi')] = \sum_{l=1}^{N} P^{(l)}(\Phi')\, Q^{(l)}(\Phi') \tag{6.3.2}$$

and homogeneous boundary conditions.

In the linearized theory the perturbation is supposed to be initially infinitesimally small and the right-hand-side of Eq. (6.3.2) is consequently neglected. One easily verifies that the linearized equation admits solutions of the form

$$\Phi'(\xi, \eta, t) = \exp[-i\,k\,\xi - \mu^{(k)}\,t]\;\varphi^{(k)}(\eta) \tag{6.3.3}$$

If, in particular, the quantity k is chosen to be a real parameter[1] then Eq. (6.3.3) represents a 'wave' of wave-length $2\pi/k$, moving in the ξ-direction with velocity $\frac{1}{k}\,\mathrm{Im}\,(\mu^{(k)})$. The amplitude of the wave grows or decays with time, depending on the sign of $Re(\mu^{(k)})$.

We shall consider shortly a somewhat different interpretation of formula (6.3.3); for the moment, let us accept in Eq. (6.3.3) a possible solution of the linearized problem.

[1] In the present study k will always be assumed real. However, another approach is also possible. Thus, Watson (1962) considers $\mu^{(k)}$ to be real, while k is complex, when studying a problem of hydrodynamic stability.

In substituting Eq. (6.3.3) into Eq. (6.3.2) the following notation will be needed:

$$\left.\begin{aligned} L[\Phi'] &= \exp[-i\,k\,\xi - \mu^{(k)}\,t]\,L_k[\varphi^{(k)}] \\ S[\Phi'] &= \exp[-i\,k\,\xi - \mu^{(k)}\,t]\,S_k[\varphi^{(k)}] \\ P^{(l)}[\Phi'] &= \exp[-i\,k\,\xi - \mu^{(k)}\,t]\,P_k^{(l)}[\varphi^{(k)}] \\ Q^{(l)}[\Phi'] &= \exp[-i\,k\,\xi - \mu^{(k)}\,t]\,Q_k^{(l)}[\varphi^{(k)}] \end{aligned}\right\} \tag{6.3.4}$$

where L_k, S_k, $P_k^{(l)}$ and $Q_k^{(l)}$ are differential operators in η, containing k as a parameter. These operators will in general be complex. In the same notation we also have

$$L(\varphi_0) = L_0(\varphi_0); \qquad P^{(l)}(\varphi_0) = P_0^{(l)}(\varphi_0); \qquad Q^{(l)}(\varphi_0) = Q_0^{(l)}(\varphi_0) \tag{6.3.5}$$

Finally, we introduce the abbreviation

$$\mathfrak{L}_k = L_k - \sum_{l=1}^{N} [Q_0^{(l)}(\varphi_0)\,P_k^{(l)} + P_0^{(l)}(\varphi_0)\,Q_k^{(l)}] \tag{6.3.6}$$

We then obtain for the function $\varphi^{(k)}$ the equation

$$(\mathfrak{L}_k + \mu^{(k)}\,S_k)\,\varphi^{(k)} = 0 \tag{6.3.7}$$

with homogeneous boundary conditions.

The problem for $\varphi^{(k)}$ is thus an eigenvalue-problem, analogous to the one discussed in section 2.3. Solutions $\varphi^{(k)} = \varphi_n^{(k)}$ of the problem exist only for eigenvalues $\mu^{(k)} = \mu_n^{(k)}$; if for all eigenvalues $Re(\mu_n^{(k)}) > 0$ then, by Eq. (6.3.3) the basic solution is stable; if for some eigenvalues $Re(\mu_n^{(k)}) < 0$, then there is instability.

As in section 2.3, we formulate certain hypotheses concerning the eigenvalues. We assume that for every value of k there exists a denumerably infinite sequence of eigenvalues $\mu_n^{(k)}$, ordered in such way that for every fixed value of k

$$Re(\mu_{n+1}^{(k)}) > Re(\mu_n^{(k)}); \qquad n = 0, 1, 2, \ldots$$

Moreover, we suppose that the operator $\mathfrak{L}_k$ contains a parameter R, characteristic of the basic solution whose stability is being studied, and that there exists a value $R = R_{cr}$ such that for $R < R_{cr}$ there is stability while for $R > R_{cr}$ there is instability. More specifically, we assume that for $R < R_{cr}$ we have $Re(\mu_n^{(k)}) > 0$ for all values of k and n, while for $R > R_{cr}$, $Re(\mu_n^{(k)}) < 0$ for at least $n = 0$, and for $k_1(R) < k < k_2(R)$. These assumptions are summarized in fig. 6.

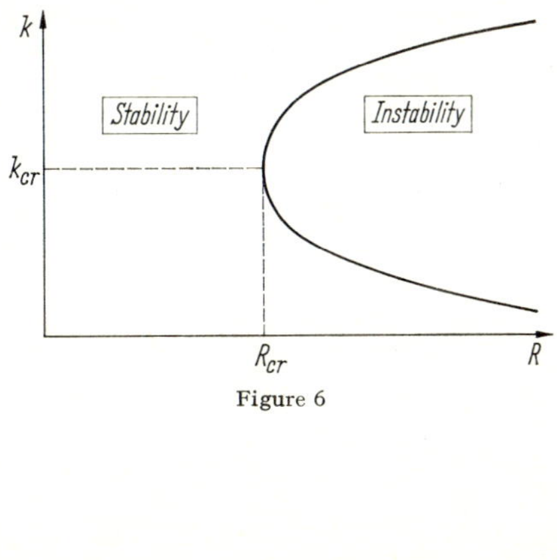

Figure 6

In the $k - R$ plane a critical curve thus separates the regions of stability and instability; as R increases from values $R < R_{cr}$, instability occurs at $R = R_{cr}$, with a wavelength $2\pi/k_{cr}$.

Since the parameter k can still take all real values, Eq. (6.3.3) only represents a particular solution of the linearized stability problem. The general solution is obtained by superposition and is given by

$$\Phi'(\xi,\eta,t) = \sum_k a^{(k)}\, e^{-ik\xi} \left\{ \sum_n e^{-\mu_n^{(k)} t} \varphi_n^{(k)}(\eta) \right\} \tag{6.3.8}$$

where the constants $a^{(k)}$ are to be determined from the initial conditions. If now $a^{(k)} \neq 0$ for all values of k, then the summation with respect to k in Eq. (6.3.8) should be replaced by a suitably defined integration, which leads to a Fourier-integral representation of the perturbation. Indeed, we could have introduced at the outset, in the linearized form of Eq. (6.3.2), a Fourier-transformation with respect to ξ. Eq. (6.3.3) would then represent a Fourier-component of the perturbation. We shall follow this approach in the analysis of the non-linear problem. However, as we shall see, the proper definition of the Fourier-transforms presents certain difficulties.

6.4 Fourier-analysis of the non-linear stability problem

In attempting to analyse the perturbation by Fourier-methods we can either assume that the integral $\int_{-\infty}^{\infty} |\Phi'|\, d\xi$ is finite and use Fourier-integrals, or assume that the perturbation is periodic in ξ and use Fourier-series. If neither one of these assumptions can be justified, then the ordinary definition of the Fourier-integral is no longer adequate and Fourier-integrals of a generalized kind have to be introduced, as explained for instance by BATCHELOR (1956). In the present analysis a less elegant, simpler method of representation proves to be more convenient.

Our aim is to represent perturbations which in general are aperiodic, but which in limiting cases may become periodic in ξ. In order to be able to cover both eventualities by the same formalism we adopt a stratagem which has frequently been used in the more primitive formulation of the theory of turbulence.

Instead of the interval $-\infty \leqq \xi \leqq +\infty$ we consider the interval $-X \leqq \xi \leqq +X$, X being very large. We write

$$\Phi'(\xi,\eta,t) = \sum_{-\infty}^{\infty} e^{-ik'\xi}\, \Phi_{k'}(\eta,t) \tag{6.4.1}$$

where $k' = \pi m/X$ and where the summation is defined with respect to all values $m = 0, \mp 1, \mp 2, \mp 3, \ldots$ Now the distance Δk between tow consecutive values of k' equals π/X, so that by choosing X sufficiently large we can make Δk very small. In the limit $X \to \infty$ the summation in Eq. (6.4.1) can be replaced by a suitably defined integral, but for the purpose of the present study, representation by a series of distinct terms is more convenient. We shall therefore utilize series (6.4.1) with the understanding that X may be chosen arbitrarily large (and thus Δk arbitrarily small), so that when considering any finite interval of the k-axis we may place in this interval as many points k' as we wish. This being the case, Eq. (6.4.1) represents periodic perturbations with period $2\pi/k_0$, where k_0 may have any finite value, if $\Phi_{k'} \equiv 0$ for all $k' \neq \mp n k_0$, $n = 0, 1, 2, \ldots$ On the other hand, if $\Phi_{k'} \neq 0$ for all values of k', the perturbation is aperiodic (in the large interval $-X \leqq \xi \leqq X$). Of course formally in Eq. (6.4.1) we always have periodicity with a very large period $2X$, but this artificial periodicity in no way affects the results in the type of problem that we consider here.

The Fourier-component $\Phi_k(\eta, t)$ is now defined by

$$\Phi_k(\eta, t) = \frac{1}{2X} \int_{-X}^{X} e^{ik\xi}\, \Phi'(\xi, \eta, t)\, d\xi \tag{6.4.2}$$

We recall that the perturbation Φ' is assumed to be real. Consequently we have

$$\Phi_{-k} = \bar{\Phi}_k \tag{6.4.3}$$

Here, and in the following pages, the quantity or operator $\bar{f}$ denotes the complex conjugate of the quantity or operator f.

In keeping with previous notation we have

$$\left.\begin{aligned}
L_k[\Phi_k] &= \frac{1}{2X} \int_{-X}^{X} e^{ik\xi}\, L[\Phi']\, d\xi \\
S_k[\Phi_k] &= \frac{1}{2X} \int_{-X}^{X} e^{ik\xi}\, S[\Phi']\, d\xi \\
P_k^{(l)}[\Phi_k] &= \frac{1}{2X} \int_{-X}^{X} e^{ik\xi}\, P^{(l)}[\Phi']\, d\xi \\
Q_k^{(l)}[\Phi_k] &= \frac{1}{2X} \int_{-X}^{X} e^{ik\xi}\, Q^{(l)}[\Phi']\, d\xi
\end{aligned}\right\} \tag{6.4.4}$$

Since the operators L, S, $P^{(l)}$ and $Q^{(l)}$ are all real, we easily find that

$$\begin{aligned}
L_{-k}[\Phi_{-k}] &= \bar{L}_k[\bar{\Phi}_k]; \qquad & S_{-k}[\Phi_{-k}] &= \bar{S}_k[\bar{\Phi}_k] \\
P_{-k}^{(l)}[\Phi_{-k}] &= \bar{P}_k^{(l)}[\bar{\Phi}_k]; \qquad & Q_{-k}^{(l)}[\Phi_{-k}] &= \bar{Q}_k^{(l)}[\bar{\Phi}_k]
\end{aligned} \tag{6.4.5}$$

Utilizing now results (6.4.1) to (6.4.5) in Eq. (6.3.2) we obtain after some calculations, the equation

$$\mathfrak{L}_k[\Phi_k] - \frac{\partial}{\partial t} S_k(\Phi_k) = \sum_{l=1}^{N} F_k^{(l)} \tag{6.4.6}$$

for every component Φ_k.

The operator $\mathfrak{L}_k$ has already been defined in Eq. (6.3.6); for $F_0^{(l)}$ we find

$$F_0^{(l)} = P_0^{(l)}(\Phi_0)\, Q_0^{(l)}(\Phi_0) + \sum_{k'=\Delta k}^{\infty} \{\bar{P}_{k'}^{(l)}[\bar{\Phi}_{k'}]\, Q_{k'}^{(l)}(\Phi_{k'}) + \bar{Q}_{k'}^{(l)}(\bar{\Phi}_{k'})\, P_{k'}^{(l)}(\Phi_{k'})\} \tag{6.4.7}$$

while $F_k^{(l)}$, $k > 0$ is given by

$$F_k^{(l)} = P_0^{(l)}(\Phi_0)\, Q_k^{(l)}(\Phi_k) + Q_0^{(l)}(\Phi_0)\, P_k^{(l)}(\Phi_k) + \sum_{k'=\Delta k}^{k-\Delta k} P_{k'}^{(l)}(\Phi_{k'})\, Q_{k-k'}^{(l)}(\Phi_{k-k'})$$

$$+ \sum_{k'=\Delta k}^{\infty} \{\bar{P}_{k'}^{(l)}(\bar{\Phi}_{k'})\, Q_{k+k'}^{(l)}(\Phi_{k+k'}) + \bar{Q}_{k'}^{(l)}(\bar{\Phi}_{k'})\, P_{k+k'}^{(l)}(\Phi_{k+k'})\} \tag{6.4.8}$$

In Eqs. (6.4.6), with definitions (6.4.7), (6.4.8), we have an infinite system of equations for the Fourier-components Φ_k. In addition, every Φ_k must satisfy homogeneous boundary conditions at $\eta = 0$ and $\eta = 1$. The terms $F_k^{(l)}$ represent the effects of interactions between components of different wave-length; in linearized approximation these terms disappear, leading to the results of section 6.3. A characteristic feature of the linearized theory is the fact that Fourier-components of different wave-length are independent of each other. In non-linear theory the interdependence of the Fourier-components is essential.

6.5 Orthogonality relations

In a formal analysis Eq. (6.4.6) can be considered as a non-homogeneous form of the equation of the linearized theory, with the right-hand-side playing the role of non-homogeneous terms. From that point of view Eq. (6.4.6) is not much different from Eq. (2.2.6), which described stability problems in one-dimensional space. We are thus led to study again eigenfunction expansions. In doing so we can use many of the results of chapter 2, but due to the somewhat more general form of Eq. (6.4.6), certain modifications are needed. These will be considered in the present section.

We define the operator $\tilde{\mathfrak{L}}_k$, adjoint of the operator $\mathfrak{L}_k$, by the relation

$$\psi_1\, \mathfrak{L}_k(\psi_2) - \psi_2\, \tilde{\mathfrak{L}}_k(\psi_1) = \frac{d}{d\eta} P(\psi_1, \psi_2) \tag{6.5.1}$$

where ψ_1 and ψ_2 are any two (suitably continuous) functions of η, and P is a bilinear concomitant. A similar relation defines $\tilde{S}_k$ as adjoint of the operator S_k.

We now consider the adjoint problem

$$[\tilde{\mathfrak{L}}_k + \lambda^{(k)} \tilde{S}_k] \, \tilde{\varphi}^{(k)} = 0 \tag{6.5.2}$$

with boundary conditions chosen so that the bilinear concomitants for $\mathfrak{L}_k$ and for S_k vanish at the boundaries $\eta = 0$ and $\eta = 1$. From Eqs. (6.5.2) and (6.3.7) after some simple manipulations we obtain the relation

$$(\lambda^{(k)} - \mu^{(k)}) \int_0^1 \tilde{\varphi}^{(k)} \, S_k(\varphi^{(k)}) \, d\eta = 0 \tag{6.5.3}$$

The eigenfunction $\varphi_n^{(k)}$ and the adjoint eigenfunction $\tilde{\varphi}_m^{(k)}$ have therefore an orthogonality relation which, after normalization, can be expressed by

$$\int_0^1 \tilde{\varphi}_n^{(k)} \, S_k(\varphi_m^{(k)}) \, d\eta \equiv \int_0^1 \varphi_m^{(k)} \, \tilde{S}_k(\tilde{\varphi}_n^{(k)}) \, d\eta = \begin{cases} 1 & \text{if} \quad n = m \\ 0 & \text{if} \quad n \neq m \end{cases} \tag{6.5.4}$$

Eq. (6.5.4) is a generalized form of the definition of bi-orthogonal functions, introduced in chapter 2.

It should perhaps be stated that the generalized orthogonality relation (6.5.4), which is fundamental for much of the analysis that follows, cannot be found in any standard text on eigenfunctions and eigenfunction expansions and appears, to this author's knowledge, not to have been formulated and utilized in the past. Indeed, the general theory of eigenfunctions for equations of the type (6.3.7) and its adjoint (6.5.2) is to date practically unexplored.

We remark that functions $\varphi_n^{(k)}$, $\tilde{\varphi}_m^{(k)}$ and operators S_k, $\tilde{S}_k$ are generally complex, so that relation (6.5.4), for $n \neq m$ implies that the real and the imaginary parts of the quantities involved vanish independently.

In the simplest case the operators S_k and $\mathfrak{L}_k$ are self-adjoint. To illustrate the orthogonality properties let us consider for example

$$S_k = \frac{\partial^2}{\partial \eta^2} - k^2; \qquad \mathfrak{L}_k = S_k^2 \tag{6.5.5}$$

We impose the boundary conditions $\varphi^{(k)}(0) = \varphi^{(k)}(1) = \varphi_\eta^{(k)}(0) = \varphi_\eta^{(k)}(1) = 0$; the eigenfunctions are then easily seen to be linear combinations of trigonometric and hyperbolic functions. It is convenient to distinguish two families of eigenfunctions: eigenfunctions that are symmetric with respect to $\eta = \frac{1}{2}$ and eigenfunctions that are antisymmetric. The symmetric family for instance is given by

$$\varphi_n^{(k)} = C \left[\frac{\cosh [k\,(\eta - \frac{1}{2})]}{\cosh \frac{1}{2} k} - \frac{\cos [\sqrt{\mu^{(k)} - k^2}\,(\eta - \frac{1}{2})]}{\cos \frac{1}{2} \sqrt{\mu^{(k)} - k^2}} \right] \tag{6.5.6}$$

where C can be considered as a normalization constant, while the eigenvalues $\mu^{(k)}$ are solutions of the equation

$$\sqrt{\mu^{(k)} - k^2} \tan \tfrac{1}{2} \sqrt{\mu^{(k)} - k^2} = -k \tanh \tfrac{1}{2} k \tag{6.5.7}$$

One verifies without difficulty that Eq. (6.5.7) defines a denumerably infinite sequence of eigenvalues.

Now the eigenfunctions $\varphi_n^{(k)}$ of Eq. (6.5.6) are *not* mutually orthogonal in the interval $0 \leqq \eta \leqq 1$, but they do satisfy the orthogonality relation (6.5.4), which in the present case takes the form

$$\int_0^1 \left[\frac{d\varphi_m^{(k)}}{d\eta} \frac{d\varphi_n^{(k)}}{d\eta} + k^2 \varphi_m^{(k)} \varphi_n^{(k)} \right] d\eta = \begin{cases} 1 & \text{if} \quad n = m \\ 0 & \text{if} \quad n \neq m \end{cases} \tag{6.5.8}$$

Returning to the general case of non-self-adjoint operators, we must remark, as in chapter 2, that further utility of the bi-orthogonal functions possessing the property (6.5.4) is subordinated to the requirement that the sequence of eigenfunctions $\varphi_n^{(k)}$ should permit to establish an expansion theorem, so that functions satisfying certain continuity conditions and the boundary conditions of the linearized problem can be expanded in the interval $0 \leqq \eta \leqq 1$ in convergent series of eigenfunctions. In what follows we shall always assume this requirement to be satisfied.

In application let us consider a non-homogeneous equation

$$[\mathfrak{L}_k + \mu S_k] U(\eta) = f(\eta) \tag{6.5.9}$$

with some homogeneous boundary conditions. $f(\eta)$ is a given function and μ a constant, not equal to any eigenvalue of the homogeneous form of Eq. (6.5.9).

We introduce the eigenfunction expansion

$$U(\eta) = \sum_{n=0}^{\infty} A_n \varphi_n^{(k)}(\eta) \tag{6.5.10}$$

Multiplying Eq. (6.5.9) by $\tilde{\varphi}_m^{(k)}$ and integrating over the interval we find

$$A_m = \frac{1}{\mu - \mu_m^{(k)}} \int_0^1 f(\eta') \tilde{\varphi}_m^{(k)}(\eta') \, d\eta' \tag{6.5.11}$$

The solution of Eq. (6.5.9) can thus be written in the form

$$U(\eta) = \int_0^1 G(\eta, \eta') f(\eta') \, d\eta' \tag{6.5.12}$$

$$G(\eta, \eta') = \sum_n \frac{\varphi_n^{(k)}(\eta) \tilde{\varphi}_n^{(k)}(\eta')}{\mu - \mu_n^{(k)}} \tag{6.5.13}$$

The function $G(\eta, \eta')$ is the Green's function for the problem.

Consider finally a somewhat more general problem, formulated by the equation

$$\left[\mathfrak{L}_k - \frac{\partial}{\partial t} S_k\right] \psi(\eta, t) = f(\eta, t) \tag{6.5.14}$$

with homogeneous boundary conditions and with $f(\eta, t)$ being a given function. By an analysis analogous to the preceding example we find that the solution is

$$\psi(\eta_0, t) = -\sum_n \left\{ \int_0^t e^{\mu_n^{(k)} t'} \left[\int_0^1 \tilde{\varphi}_n^{(k)} f(\eta, t')\, d\eta \right] dt' + C_n \right\} e^{-\mu_n^{(k)} t}\, \varphi_n^{(k)}(\eta_0) \tag{6.5.15}$$

where C_n are constants to be determined from the initial conditions.

In the analysis that follows eigenfunction expansions leading to solutions of the type (6.5.15) will frequently be used. It should however be kept in mind that in the absence of a satisfactory general theory for eigenvalue problems defined by Eq. (6.3.7) our analysis is in a sense formal, that is valid in those cases in which certain hypotheses are verified. Summarizing let us state once again that we have assumed that Eq. (6.3.7) has a denumerable infinite sequence of eigenvalues, and that the corresponding eigenfunctions form a set of functions for which an expansion theorem holds, so that, under suitable continuity conditions, convergent series expansions in the interval $0 \leqq \eta \leqq 1$ can be defined.

6.6 Initial conditions

In the classical formulation, the stability problem is considered as an initial value problem. It is supposed that at the initial instant $t = 0$ a known perturbation $\Phi'(\xi, \eta, 0)$ is introduced. Using Eq. (6.4.2) the initial values of the Fourier-components Φ_k are then obtained. We have already seen that this formulation is somewhat artificial and that it is desirable to allow for different perturbations to be introduced in arbitrary order during the time history of the phenomenon. As in the case of the one-dimensional problems we can cover this eventuality by studying not specific initial values, but classes of initial values, defined by bounds on the orders of magnitude of the perturbations. If we assume that Φ_k is initialy limited by a certain order of magnitude requirement (the order of magnitude imposed varying of course with the parameter k) and if we find that solutions of Eqs. (6.4.6) are at most of this order of magnitude throughout the time history, then it makes no difference when and how many times different perturbations are introduced, as long as at all times the order of magnitude requirement is respected. In fact we study the stability of the basic solution with respect to perturbations of a certain prescribed level.

In general the perturbations will be such that $\Phi_k \neq 0$ for all values of k. However, a very special and much more restrictive distribution of Φ_k also satisfies the system of Eqs. (6.4.6), namely the distribution

$$\left.\begin{aligned} \Phi_k &= 0 \quad \text{for} \quad k \neq m\, k_0 \\ \Phi_k &\neq 0 \quad \text{for} \quad k = m\, k_0 \end{aligned}\right\} \quad m = 0, 1, 2, \ldots \tag{6.6.1}$$

where k_0 has a fixed finite value. We shall speak in this case of a periodic solution. Clearly, by Eq. (6.4.1), the perturbation is periodic in ξ, with a period $2\pi/k_0$.

Formally, when studying periodic solutions, we must suppose that the initial values are periodic too, and that at no instant of time, during the time history of phenomenon, can an aperiodic perturbation be introduced. This may seem to restrict the scope of applications very severely. Indeed, in chapter 1, we have already concluded that in order to represent the realities of the physical world we should consider perturbations as entirely arbitrary functions of the space variables. Nevertheless, the periodic solutions appear on closer analysis to be of considerable interest, not only from mathematical point of view, but also as solutions of the general problem of stability.

Intuitively there are some reasons to expect that under certain conditions periodic solutions may appear in the limit as $t \to \infty$ for certain aperiodic initial perturbations. If, in fact, we consider in fig. 6 the case of weak instability, that is when R is close to R_{cr}, $(R - R_{cr}) > 0$, then in the linearized theory, for any initial values, almost all Fourier-components tend to zero as $t \to \infty$, except for the components of wave-length close to $2\pi/k_{cr}$, whose amplitude grows without bounds. It would not be too surprising if in non-linear analysis the predominant role of the wave-length $2\pi/k_{cr}$ were retained, leading to a periodic solution of this wave-length.

Of course, the intuitive argument is not sufficient to claim an important role for periodic solutions in the theory of stability. As we shall see below, however, the expectation formulated above is borne out at least to some degree by a more precise mathematical analysis.

In the next chapter we shall develop an asymptotic theory of the periodic solutions of Eqs. (6.4.6), these solutions being considered as a formal consequence of periodic initial values. In chapter 8 we shall return to the more general (and more realistic) problem of stability and consider perturbations that are aperiodic. We shall then demonstrate that, under certain conditions, *some* periodic solutions represent the limit for $t \to \infty$ of certain classes of initialy aperiodic perturbations.

Chapter 7

ASYMPTOTIC THEORY OF PERIODIC SOLUTIONS

7.1 Basic equations and transformations

When studying periodic solutions of a period $2\pi/k_0$ we have $\Phi_k = 0$ for $k \neq m\,k_0$. In order to simplify the notation we shall write in this case $\Phi_{m k_0} = \Phi_m$, $\mathfrak{L}_{m k_0} = \mathfrak{L}_m$, etc. Eqs. (6.4.6) to (6.4.8) then take the form

$$\mathfrak{L}_m[\Phi_m] - \frac{\partial}{\partial t} S_m[\Phi_m] = \sum_{l=1}^{N} F_m^{(l)}, \tag{7.1.1}$$

$$F_0^{(l)} = P_0^{(l)}(\Phi_0)\, Q_0^{(l)}(\Phi_0) + \sum_{n=1}^{\infty} \{\bar{P}_n^{(l)}(\bar{\Phi}_n)\, Q_n^{(l)}(\Phi_n) + \bar{Q}_n^{(l)}(\bar{\Phi}_n)\, P_n^{(l)}(\Phi_n)\}, \tag{7.1.2}$$

$$F_m^{(l)} = P_0^{(l)}(\Phi_0)\, Q_m^{(l)}(\Phi_m) + Q_0^{(l)}(\Phi_0)\, P_m^{(l)}(\Phi_m) + \sum_{n=1}^{m-1} P_m^{(l)}(\Phi_m)\, Q_{m-n}^{(l)}(\Phi_{m-n})$$
$$+ \sum_{n=1}^{\infty} \{\bar{P}_n^{(l)}(\bar{\Phi}_n)\, Q_{m+n}^{(l)}(\Phi_{m+n}) + \bar{Q}_n^{(l)}(\bar{\Phi}_n)\, P_{m+n}^{(l)}(\Phi_{m+n})\}\,; \quad (m \neq 0). \tag{7.1.3}$$

with the understanding that the finite sum in Eq. (7.1.3) is zero when $m \leqq 1$.

We consider problems of weak stability or instability, characterized by the fact that in fig. 6 the point (k_0, R) is situated close to the critical curve. More precisely, we shall suppose that for $m \neq 1$ $Re(\mu_p^{(m)}) > 0$ for all values of p, while $Re(\mu_p^{(1)}) > 0$ for $p \neq 0$. Thus one and only one eigenvalue $\mu_0^{(1)}$ can possibly have a negative real part. In accordance with our earlier definition, we have weak instability if $Re(\mu_0^{(1)}) \to 0$ through negative values and weak stability when $Re(\mu_0^{(1)}) \to 0$ through positive values. We remark that although k_0 does not necessarily equal k_{cr}, the definition of weak stability and instability implies that k_0 is restricted to some neighbourhood of k_{cr}, since otherwise the real part of eigenvalues different from $\mu_0^{(1)}$ may approach zero. The exact extent of the admissible k_0-values depends on the form of the critical curve.

Our aim is to construct asymptotic solutions with respect to suitably defined small parameters. In this we shall follow and generalize the methods that we have employed in the study of problems in one-dimensional space. We begin by introducing a formal transformation

$$\Phi_m = \varepsilon\, \delta_m\, \psi_m; \qquad \bar{\Phi}_m = \varepsilon\, \delta_m\, \bar{\psi}_m \tag{7.1.4}$$

We then have

$$\mathfrak{L}_m(\psi_m) - \frac{\partial}{\partial t} S_m(\psi_m) = \sum_{l=1}^{N} F_m^{(l)} \tag{7.1.5}$$

where now:

$$F_0^{(l)} = \varepsilon\,\delta_0\, P_0^{(l)}(\psi_0)\, Q_0^{(l)}(\psi_0) + \frac{\varepsilon}{\delta_0} \sum_{n=1}^{\infty} \delta_n^2 \{\bar{P}_n^{(l)}(\bar{\psi}_n)\, Q_n^{(l)}(\psi_n) + \bar{Q}_n^{(l)}(\bar{\psi}_n)\, P_n^{(l)}(\psi_n)\}, \tag{7.1.6}$$

$$\begin{aligned} F_m^{(l)} &= \varepsilon\,\delta_0\, \{P_0^{(l)}(\psi_0)\, Q_m^{(l)}(\psi_m) + Q_0^{(l)}(\psi_0)\, P_m^{(l)}(\psi_m)\} + \\ &+ \frac{\varepsilon}{\delta_m} \sum_{n=1}^{m-1} \delta_n\, \delta_{m-n}\, P_n^{(l)}(\psi_n)\, Q_{m-n}^{(l)}(\psi_{m-n}) + \\ &+ \frac{\varepsilon}{\delta_m} \sum_{n=1}^{\infty} \delta_n\, \delta_{m+n} \{\bar{P}_n^{(l)}(\bar{\psi}_n)\, Q_{m+n}^{(l)}(\psi_{m+n}) + \bar{Q}_n^{(l)}(\bar{\psi}_n)\, P_{m+n}^{(l)}(\psi_{m+n})\}; \\ & \qquad (m \neq 0) \end{aligned} \tag{7.1.7}$$

The role of the parameters ε and δ_m is to measure the order of magnitude of the components Φ_m. We shall hence attempt to choose these parameters in such way that ψ_m be of order of magnitude unity. As in chapter 4, the transformation (7.1.4) is meant to give an a priori estimate of the order of magnitude of the solutions.

We can choose for ε the order of magnitude of the largest component Φ_m; in that case we shall have $|\delta_m| \leqq 1$ and ε can be interpreted as the level of perturbations. We shall suppose that perturbations are small, so that $|\varepsilon| \ll 1$, and seek asymptotic expansions of the solutions of Eqs. (7.1.5) valid for $\varepsilon \to 0$.

The parameters ε and δ_m must of course be expressed in terms of the natural small parameter for the problem of weak stability or instability, that is in terms of $Re(\mu_0^{(1)})$. In the analysis which follows we shall first use the concept of forced solutions and supposing ε small, determine the relation between ε and δ_m. Finally the relation between ε and $Re(\mu_0^{(1)})$ will be established.

7.2 Forced solutions for the components Ψ_m, $m \neq 1$

Eq. (7.1.5) can be considered as a non-homogeneous equation of the type studied in section 6.5. Using the result (6.5.15), we can write

$$\psi_m(\eta, t) = \sum_{p=0}^{\infty} A_p^{(m)}(t)\, \varphi_p^{(m)}(\eta) \tag{7.2.1}$$

$$A_p^{(m)}(t) = -e^{-\mu_p^{(m)} t} \left\{ \int^{t} e^{\mu_p^{(m)} t'}\, b_p^{(m)}(t')\, dt' + C_p^{(m)} \right\} \tag{7.2.2}$$

where

$$b_p^{(m)} = \int_0^1 \sum_{l=1}^{N} F_m^{(l)} \tilde{\varphi}_p^{(m)}(\eta)\, d\eta \tag{7.2.3}$$

and where $C_p^{(m)}$ are constants, to be determined from initial conditions.

Since we have supposed $Re(\mu_p^{(m)}) > 0$ for $m \neq 1$, the term $C_p^{(m)} \exp[-\mu_p^{(m)} t]$ decays in time and tends to zero when $t \to \infty$. The first term on the right-hand-side of Eq. (7.2.2) remains bounded if $b_p^{(m)}$ is bounded as a function of time.

It follows that the order of magnitude of $A_p^{(m)}$, for $\varepsilon \to 0$, is determined by the order of magnitude of $b_p^{(m)}$ and hence by the order of magnitude of $F_m^{(l)}$. If $\psi_m(\eta, t)$ is to be of order of magnitude unity for $\varepsilon \to 0$, $A_p^{(m)}$ and hence $F_m^{(l)}$ must be of order of magnitude unity. Looking over the definition of $F_m^{(l)}$, Eq. (7.1.6), (7.1.7), we see that $F_m^{(l)} = O(1)$ if $\delta_m = O(\varepsilon)$ for $m \neq 1$. Consequently, we can write

$$\delta_1 = 1 \tag{7.2.4}$$

Analysing Eq. (7.1.6) and (7.1.7) more precisely we find that *all* $F_m^{(l)}$, $m \neq 1$, are of the order of magnitude unity if we choose

$$\delta_m = \varepsilon^{|m-1|} \tag{7.2.5}$$

Introducing this results into Eqs. (7.1.6), (7.1.7) we further find, in the first asymptotic approximation,

$$F_0^{(l)} = \bar{P}_1^{(l)}(\bar{\psi})\, Q_1^{(l)}(\psi_1) + \bar{Q}_1^{(l)}(\bar{\psi}_1)\, P_1^{(l)}(\psi_1) + O(\varepsilon^2) \tag{7.2.6}$$

$$F_m^{(l)} = \sum_{n=1}^{m-1} P_n^{(l)}(\psi_n)\, Q_{m-n}^{(l)}(\psi_{m-n}) + O(\varepsilon^2); \quad \text{for } m > 1 \tag{7.2.7}$$

Thus, for small perturbations, that is for $\varepsilon \ll 1$, $F_0^{(l)}$ only depends upon ψ_1, while $F_m^{(l)}$ for $m > 1$ only depend upon ψ_n, $n < m$. If $\psi_1$is known, ψ_0 and ψ_2 can immediately be calculated. Next ψ_3 can be expressed in terms of ψ_0, ψ_1, ψ_2; etc. All components ψ_m, $m \neq 1$, can thus be expressed in terms of ψ_1. Furthermore, from Eq. (7.2.2) it follows that the components ψ_m, $m \neq 1$, take values different from zero in the limit $t \to \infty$ only if ψ_1 differs from zero. The asymptotic solutions defined above are therefore forced solutions, analogous to those considered in chapter 3.

In interpreting these forced solutions it is useful to consider separately ψ_0 and ψ_m, $m \geqq 2$. The components ψ_m ($m \geqq 2$) are the harmonics of ψ_1, generated by non-linear interactions; ψ_0 is of somewhat different character. We note that if in the limit $t \to \infty$ ψ_1 approaches a stationary value or a value harmonic in time, then $F_0^{(l)}$ becomes time-independent and ψ_0 approaches a stationary value. Returning to the Fourier-decomposition of the perturbations by formula (6.4.1) we see that in that case the perturbation contains a component independent of ξ and t. Now the basic solution, whose stability is studied, is defined as the solution of the original problem which is independent of ξ and t. Thus, in presence of perturbations, a modification of the basic solution occurs, the modification being expressed by the term ψ_0. This effect is sometimes called 'the effect on the mean'. It is also present if ψ_1 is not purely harmonic, but is for instance periodic in time. ψ_0 then contains a stationary modification of the basic solution and some terms which are oscillatory in time.

7.3 Analysis of the component Ψ_1

We now turn to Eqs. (7.1.5), (7.1.7) for $m = 1$. Using again the results of section 6.5 we can write

$$\psi_1(\eta, t) = \sum_{p=0}^{\infty} A_p^{(1)}(t)\, \varphi_p^{(1)}(\eta) \tag{7.3.1}$$

$$\frac{d A_p^{(1)}}{dt} + \mu_p^{(1)} A_p^{(1)} = -\varepsilon^2 b_p^{(1)} \tag{7.3.2}$$

$$b_p^{(1)} = \frac{1}{\varepsilon^2} \sum_{l=1}^{N} \int_0^1 F_1^{(l)}\, \tilde{\varphi}_p^{(1)}(\eta)\, d\eta \tag{7.3.3}$$

Furthermore, δ_m being defined by Eq. (7.2.5), we have for $F_1^{(l)}$ the following result:

$$\begin{aligned} F_1^{(l)} = \varepsilon^2 \{ & P_0^{(l)}(\psi_0)\, Q_1^{(l)}(\psi_1) + Q_0^{(l)}(\psi_0)\, P_1^{(l)}(\psi_1) + \\ & + \bar{P}_1^{(l)}(\bar{\psi}_1)\, Q_2^{(l)}(\psi_2) + \bar{Q}_1^{(l)}(\bar{\psi}_1)\, P_2^{(l)}(\psi_2)\} + O(\varepsilon^4) \end{aligned} \tag{7.3.4}$$

It is seen that, in the first asymptotic approximation, only the components ψ_0 and ψ_2 interact directly with ψ_1. Since $\psi_n = O(1)$, we have $F_1^{(l)} = O(\varepsilon^2)$ and $b_p^{(1)} = O(1)$. Let us presently write

$$\mu_0^{(1)} = i\, \nu_0 + \tau_0 \tag{7.3.5}$$

For weak stability or instability $\tau_0 \to 0$. Consider Eq. (7.3.2) for $p = 0$. After some simple manipulations we can deduce an equation for $|A_0^{(1)}|$, which reads as follows:

$$\frac{d |A_0^{(1)}|^2}{dt} + 2\tau_0\, |A_0^{(1)}|^2 = -\varepsilon^2 \{\bar{A}_0^{(1)} b_p^{(1)} + A_0^{(1)} \bar{b}_p^{(1)}\} \tag{7.3.6}$$

Eq. (7.3.6) suggests that we should choose

$$\varepsilon^2 = |\tau_0| \tag{7.3.7}$$

The asymptotic expansion $\varepsilon \to 0$ is then a development for weak weak stability and instability. On the other hand, in Eq. (7.3.6), after a simple change of the time-scale, an equation is obtained in which all coefficients are of order of magnitude unity, so that solutions of order of magnitude unity may be expected.

Turning now to $A_p^{(1)}$, $p > 0$, from Eq. (7.3.2) it follows that

$$\frac{d A_p^{(1)}}{dt} + \mu_p^{(1)} A_p^{(1)} = 0(\varepsilon^2) \tag{7.3.8}$$

Now, for $p > 0$, $Re(\mu_p^{(1)})$ is positive and of order of magnitude unity. Consequently, we may write

$$A_p^{(1)} = C_p^{(1)} e^{-\mu_p^{(1)} t} + 0(\varepsilon^2) \tag{7.3.9}$$

In other words, $A_p^{(1)}$ for $p > 0$ consists of a term which decays to zero as $t \to \infty$, and a term of the order of ε^2. We shall show later on that the term of order ε^2 in Eq. (7.3.9) is a forced solution, which can be expressed as a function of $A_0^{(1)}$. For the moment we write, in accordance with Eqs. (7.3.1) and (7.3.9),

$$\psi_1(\eta, t) = A_0^{(1)}(t)\, \varphi_0^{(1)}(\eta) + 0(\varepsilon^2) \tag{7.3.10}$$

The amplitude function $A_0^{(1)}$ is a solution of Eq. (7.3.2) for $p = 0$, however, the right-hand-side of this equation still contains the components ψ_0 and ψ_2. In section 7.2 we have established the forced solutions for these two components, they can be expressed in terms of ψ_1 and hence in terms of $A_0^{(1)}$. We now proceed to develop these relations.

7.4 Further analysis of the forced solutions for Ψ_m, $m \neq 1$

Returning to the analysis of section 7.2 we introduce the result (7.3.10) into Eqs. (7.2.6) and (7.2.7). We obtain:

$$F_0^{(l)}(\eta, t) = |A_0^{(1)}|^2 F_0^{(l)}(\eta) + 0(\varepsilon^2) \tag{7.4.1}$$

$$f_0^{(l)}(\eta) = \bar{P}_1^{(l)}(\bar{\varphi}_0^{(1)})\, Q_1^{(l)}(\varphi_0^{(1)}) + \bar{Q}_1^{(l)}(\bar{\varphi}_0^{(1)})\, P_1^{(l)}(\varphi_0^{(1)})$$

$$F_2^{(l)}(\eta, t) = [A_0^{(1)}]^2 f_2^{(l)}(\eta) + 0(\varepsilon^2); \quad f_2^{(l)}(\eta) = P_1^{(l)}(\varphi_0^{(1)})\, Q_1^{(l)}(\varphi_0^{(1)}) \tag{7.4.2}$$

Thus, in the asymptotic approximation, $F_0^{(l)}$ and $F_2^{(l)}$ are products of functions of time and functions of η. We proceed to consider the forced solution for ψ_0 in greater detail. Introducing result (7.4.1) into Eqs. (7.2.1) to (7.2.4) we find

$$\psi_0(\eta, t) = \sum_{p=0}^{\infty} A_p^{(0)}(t)\, \varphi_p^{(0)}(\eta)$$

$$A_p^{(0)}(t) = -e^{-\mu_p^{(0)} t} \left\{ \int^{t} |A_0^{(1)}|^2 e^{\mu_p^{(0)} t'} dt' \int_0^1 \sum_{l=1}^{N} f_0^{(l)}\, \tilde{\varphi}_p^{(0)}(\eta)\, d\eta + C_p^{(0)} \right\} \tag{7.4.3}$$

The integral with respect to time in Eq. (7.4.3) can be further simplified by performing an integration by parts, as follows:

$$\int^{t} |A_0^{(1)}|^2 e^{\mu_p^{(0)} t'} dt' = \frac{1}{\mu_p^{(0)}} \left\{ e^{\mu_p^{(0)} t} |A_0^{(1)}|^2 - \int^{t} \frac{d|A_0^{(1)}|^2}{dt'} e^{\mu_p^{(0)} t'} dt' \right\} \qquad (7.4.4)$$

Now, from Eq. (7.3.6) we have

$$\frac{d|A_0^{(1)}|^2}{dt} = 0(\varepsilon^2) \qquad (7.4.5)$$

Consequently

$$\int^{t} |A_0^{(1)}|^2 e^{\mu_p^{(0)} t'} dt' = \frac{1}{\mu_p^{(0)}} e^{\mu_p^{(0)} t} |A_0^{(1)}|^2 + 0(\varepsilon^2) \qquad (7.4.6)$$

The forced solution for ψ_0 can thus be written in the following form

$$\psi_0(\eta, t) = |A_0^{(1)}|^2 g_0(\eta) + 0(\varepsilon^2) \qquad (7.4.7)$$

where the function g_0 is defined by the eigenfunction expansion

$$g_0(\eta) = -\sum_{p=0}^{\infty} \left[\int_0^1 \sum_{l=1}^{N} f_0^{(l)} \tilde{\varphi}_p^{(0)}(\eta') \, d\eta' \right] \frac{\varphi_p^{(0)}(\eta)}{\mu_p^{(0)}} \qquad (7.4.8)$$

The series (7.4.8) may not always be the most convenient form for the effective calculation of g_0. Alternatively, this function can be defined as the solution of the non-homogeneous ordinary differential equation

$$\mathfrak{L}_0(g_0) = \sum_{l=1}^{N} f_0^{(l)} \qquad (7.4.9)$$

Definitions (7.4.8) and (7.4.9) are easily shown to be equivalent, but a calculation starting from Eq. (7.4.9) may be more convenient if the independent solutions of the homogeneous form of (7.4.9) are known. Of course, if definition (7.4.9) is used, the proper boundary conditions for ψ_0 have yet to be imposed.

Turning to ψ_2, we can perform an entirely analogous analysis. Omitting the details we state the results. ψ_2 can be shown to be

$$\psi_2(\eta, t) = [A_0^{(1)}]^2 g_2(\eta) + 0(\varepsilon^2) \qquad (7.4.10)$$

the function g_2 being defined by the eigenfunction expansion

$$g_2(\eta) = -\sum_{p=0}^{\infty} \left[\int_0^1 \sum_{l=1}^{N} f_2^{(l)}(\eta') \, \tilde{\varphi}_p^{(2)}(\eta') \, d\eta' \right] \frac{\varphi_p^{(2)}(\eta)}{\mu_p^{(2)} - 2i\nu_0} \qquad (7.4.11)$$

g_2 can alternatively be defined as the solution of the ordinary non-homogeneous differential equation

$$[\mathfrak{L}_2 + 2i\,\nu_0\,S_2]\,g_2 = \sum_{l=1}^{N} f_2^{(l)}(\eta) \tag{7.4.12}$$

which satisfies the boundary conditions imposed on ψ_2.

7.5 The equations of the asymptotic approximation

We now summarize the results of the preceding section and bring them into a final form. We recall that $\mu_p^{(m)}$ are the eigenvalues and $\varphi_p^{(m)}$ the eigenfunctions of the linearized problem for the component Φ_m, which are considered known in the non-linear analysis.

We study the case of weak stability or instability, characterized by

$$\mu_0^{(1)} = i\,\nu_0 + \tau_0; \qquad |\tau_0| \to 0 \tag{7.5.1}$$

We have introduced the transformation

$$\Phi_m = \varepsilon\,\delta_m\,\psi_m \tag{7.5.2}$$

$$\delta_m = \varepsilon^{|m-1|}; \qquad \varepsilon^2 = |\tau_0| \tag{7.5.3}$$

and we have found that

$$\psi_0(\eta, t) = |A_0^{(1)}(t)|^2\,g_0(\eta) + 0(\varepsilon^2) \tag{7.5.4}$$

$$\psi_1(\eta, t) = \sum_{n=0}^{\infty} A_n^{(1)}(t)\,\varphi_n^{(1)}(\eta) \tag{7.5.5}$$

$$\psi_2(\eta, t) = [A_0^{(1)}(t)]^2\,g_2(\eta) + 0(\varepsilon^2) \tag{7.5.6}$$

For the purpose of further analysis the functions g_0 and g_2 may be considered known (they have been determined in section 7.4). Furthermore, all components ψ_m, $m > 2$, can be determined following section 7.2, once the solutions for ψ_0, ψ_1 and ψ_2 have been obtained. The problem is thus reduced to the study of the component ψ_1.

Taking up the results of section 7.3 and introducing for ψ_0 and ψ_2 the forced solutions (7.5.4) and (7.5.6) we find that $A_0^{(1)}$ must be a solution of the equation

$$\frac{d A_0^{(1)}}{dt} + \mu_0^{(1)}\,A_0^{(1)} = -|\tau_0|\,\beta_0\,A_0^{(1)}\,|A_0^{(1)}|^2 + 0(\varepsilon^4) \tag{7.5.7}$$

where

$$\beta_0 = \sum_{l=1}^{N} \int_0^1 f_1^{(l)}(\eta)\,\bar{\varphi}_0^{(1)}(\eta)\,d\eta \tag{7.5.8}$$

$$f_1^{(l)} = P_0^{(l)}(g_0)\,Q_1^{(l)}(\varphi_0^{(1)}) + Q_0^{(l)}(g_0)\,P_1^{(l)}(\varphi_0^{(1)}) + P_2^{(l)}(g_2)\,\bar{Q}_1^{(l)}(\bar{\varphi}_0^{(1)}) + Q_2^{(l)}(g_2)\,\bar{P}_1^{(l)}(\bar{\varphi}_0^{(1)}) \tag{7.5.9}$$

Having determined the constant β_0 the behaviour of $A_0^{(1)}$ can be analysed, and this we shall do in section 7.6. We note here, however, that in deducing Eq. (7.5.7) we have substituted in Eq. (7.3.4) the forced solutions for ψ_0 and ψ_2 *only*. Thus we have neglected the influence of the exponentially decaying terms of ψ_0 and ψ_2, such as $C_p^{(0)} \exp[-\mu_p^{(0)} t]$, etc. This procedure needs a justification, which can be obtained from the fact that the time-scale of these terms is given by $Re(\mu_p^{(0)})$ while the time-scale of $|A_0^{(1)}|$ is given by $Re[\mu_0^{(1)}]$. Now $Re[\mu_p^{(0)}] = 0(1)$ while $Re[\mu_0^{(1)}] = 0(\varepsilon^2)$, so that in the asymptotic approximation the exponential terms decay to zero very much faster than $|A_0^{(1)}|$ does change with time. To illustrate this we consider the following simple example

$$\frac{d|A_0^{(1)}|^2}{dt} + 2\tau_0 |A_0^{(1)}|^2 = -2|\tau_0||A_0^{(1)}|^2[Re(\beta_0)\,|A_0^{(1)}|^2 + C_m e^{-\tau t}] \quad (7.5.10)$$

where C_m is a constant and where τ is an arbitrary positive number of order unity. We introduce a transformation of the time-scale

$$|\tau_0|\, t = t^* \quad (7.5.11)$$

Eq. (7.5.10) then becomes

$$\frac{d|A_0^{(1)}|^2}{dt^*} + 2\frac{\tau_0}{|\tau_0|}|A_0^{(1)}|^2 = -2|A_0^{(1)}|^2\left[Re(\beta_0)\,|A_0^{(1)}|^2 + C_m e^{-\frac{\tau}{|\tau_0|}t^*}\right] \quad (7.5.12)$$

Clearly, in the asymptotic approximation for $\tau_0 \to 0$ the exponential term vanishes for every $t^* > 0$. A complete justification of Eq. (7.5.7) can be obtained along these lines, but the details of the calculation are somewhat lengthy.

Turning finally to $A_p^{(1)}$, $p > 0$, one finds without difficulty

$$A_p^{(1)}(t) = e^{-\mu_p^{(1)} t}\left\{\varepsilon^2 \beta_p \int^t e^{\mu_p^{(1)} t'}\, |A_0^{(1)}|^2 A_0^{(1)}\, dt' + C_p^{(1)}\right\} \quad (7.5.13)$$

$$\beta_p = \sum_{l=1}^{N} \int_0^1 f_1^{(l)}(\eta)\, \tilde{\varphi}_p^{(1)}(\eta)\, d\eta \quad (7.5.14)$$

In Eqs. (7.5.13), (7.5.14) we have an explicit form of the result (7,3.9) obtained earlier, and we see that in the limit $t \to \infty$, $A_p^{(1)}$, $p > 0$, approaches a forced solution of order ε^2.

The final step in the analysis of the periodic solution is the determination of the amplitude function $A_0^{(1)}$. That function is a solution of Eq. (7.5.7), of which all coefficients are known quantities. In particular, the coefficient β_0, which summarizes the non-linear effects on $A_0^{(1)}$ in the asymptotic approximation, has been expressed in terms of functions known from the linearized theory and may therefore be considered a given quantity in further analysis.

When $A_0^{(1)}$ has been determined, all other quantities, such as ψ_m for $m \neq 1$ and $A_p^{(1)}$ for $p > 0$, can be evaluated. However, a posteriori we must check that the solutions obtained are consistent with the basic hypothesis of the asymptotic expansion, namely, that all components ψ_m are of order of magnitude unity as $\varepsilon \to 0$.

7.6 Harmonic solutions

Eq. (7.5.7) is sufficiently simple to be integrated directly. We shall not go into the details of the calculation, which is easily performed; instead we follow the approach used in the study of one-dimensional problems: we first determine the permanent-type solutions, and then investigate the conditions under which the solutions obtained represent the limit of perturbations.

We note that the complete exact system of equations (7.1.1), (7.1.2), (7.1.3) admits solutions whose behaviour in time is given by

$$\Phi_m(\eta, t) = e^{i\omega m t}\, \overset{\infty}{\Phi}_m(\eta) \tag{7.6.1}$$

where ω is a real quantity. We now analyse these harmonic solutions in the asymptotic approximation. Thus, we write

$$A_0^{(1)}(t) = e^{i\omega t}\, a_0^{(1)} \tag{7.6.2}$$

where $a_0^{(1)}$ is a constant. From Eqs. (7.5.4), (7.5.6) we have

$$\psi_0(\eta, t) = |a_0^{(1)}|^2\, g_0(\eta) + 0(\varepsilon^2) \tag{7.6.3}$$

$$\psi_2(\eta, t) = e^{2i\omega t}[a_0^{(1)}]^2\, g_2(\eta) + 0(\varepsilon^2) \tag{7.6.4}$$

etc.

Eq. (7.5.7) leads to

$$i\,\omega + \mu_0^{(1)} = -|\tau_0|\,\beta_0\,|a_0^{(1)}|^2 + 0(\varepsilon^4) \tag{7.6.5}$$

Separating the real and imaginary parts we finally find

$$|a_0^{(1)}|^2 = -\frac{\tau_0}{|\tau_0|}\,\frac{1}{Re\,(\beta_0)} + 0(\varepsilon^2) \tag{7.6.6}$$

$$\omega = -\nu_0 - |\tau_0|\,\mathrm{Im}\,(\beta_0)\,|a_0^{(1)}|^2 + 0(\varepsilon^4) \tag{7.6.7}$$

Eq. (7.6.6) determines the amplitude of the harmonic solution in the asymptotic approximation. We see that this solution is indeed of order of magnitude unity if $Re(\beta_0) \neq 0$, as $\varepsilon \to 0$. Eq. (7.6.7) determines the 'wave-velocity' (see section 6.3) and contains a non-linear correction of the linearized result.

Now solution of Eq. (7.6.6) exists only under certain conditions: for weak instability ($\tau_0 < 0$) the harmonic solution exists if $Re(\beta_0) > 0$; for weak stability ($\tau_0 > 0$) the solution exists if $Re(\beta_0) < 0$.

On the other hand, the condition $Re(\beta_0) \neq 0$ for $\varepsilon \to 0$ is essential for the validity of the solution. We conclude that in any particular problem in which that condition is satisfied and the present analysis applies, the harmonic solution exists either for weak instability or for weak stability, depending on the sign of $Re(\beta_0)$.

Working backwards, the components ψ_m, $m \neq 1$, and the amplitude functions $A_p^{(1)}$, $p > 0$, can now be calculated, but the explicit formulas do not present any particular interest. Instead, we investigate the interpretation to be given to the harmonic solutions obtained above, that is, we analyse the conditions under which these solutions represent the limit of perturbations as $t \to \infty$.

From Eq. (7.5.7) we deduce the equation governing $|A_0^{(1)}|$, which reads

$$\frac{d\,|A_0^{(1)}|^2}{dt} + 2\tau_0\,|A_0^{(1)}|^2 = -2\,|\tau_0|\,Re(\beta_0)\,|A_0^{(1)}|^4 + O(\varepsilon^4) \qquad (7.6.8)$$

Utilizing result (7.6.6) we have

$$\frac{d\,|A_0^{(1)}|^2}{dt} = 2\tau_0\,|A_0^{(1)}|^2 \left\{ \frac{|A_0^{(1)}|^2}{|a_0^{(1)}|^2} - 1 \right\} + O(\varepsilon^4) \qquad (7.6.9)$$

The behaviour of $|A_0^{(1)}|$ as a function of time can now be analysed in the phase-plane, in a manner similar to the investigations of chapter 3. First we consider weak instability, $\tau_0 < 0$; when the harmonic solution $a_0^{(1)}$ exists the result is indicated in fig. 7.

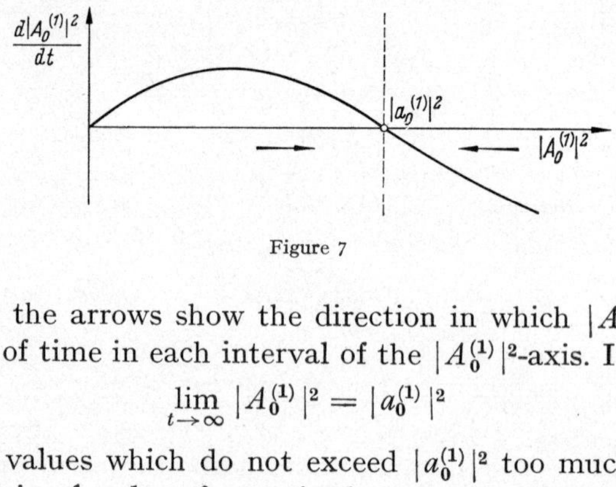

Figure 7

As before, the arrows show the direction in which $|A_0^{(1)}|^2$ develops as a function of time in each interval of the $|A_0^{(1)}|^2$-axis. It follows that

$$\lim_{t\to\infty} |A_0^{(1)}|^2 = |a_0^{(1)}|^2 \qquad (7.6.10)$$

for all initial values which do not exceed $|a_0^{(1)}|^2$ too much (that is, as long as $|A_0^{(1)}|$ is of order of magnitude unity, so that the analysis is valid).

Turning to the case of weak stability, if the harmonic solution exists, we obtain in the phase-plane the result indicated in fig. 8.

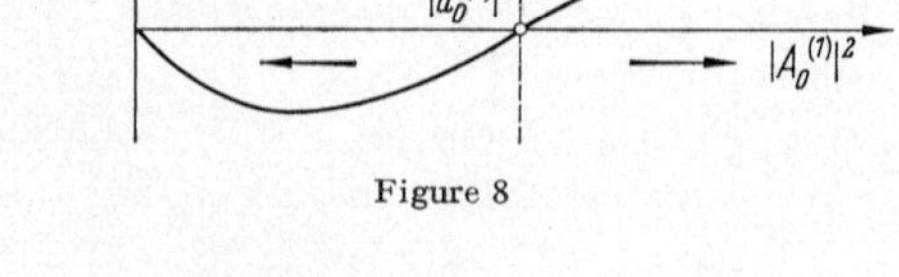

Figure 8

Clearly, for initial values smaller than $|a_0^{(1)}|$ the perturbation tends to zero as $t \to \infty$.

For initial values exceeding $|a_0^{(1)}|$ the perturbation grows as a function of time. In that case the asymptotic solution is valid only during some initial time interval; if a limiting value for $|A_0^{(1)}|$ exists, then it must be of an order of magnitude larger than unity.

In the case $\tau_0 > 0$ the harmonic solution of amplitude $|a_0^{(1)}|$ should hence be interpreted as a critical value for perturbations: there is stability with respect to perturbations smaller than $|a_0^{(1)}|$, and instability with respect to perturbations whose amplitude exceeds that critical value.

Summarizing, we can state that if the harmonic solution exists for weak instability, it represents the limit of perturbations for $t \to \infty$. If the harmonic solution exists for weak stability then it represents a critical value for instability.

We recall that the essential condition for the validity of the analysis is that in Eq. (7.5.2) the components ψ_m be of order unity. We can now generalize the results beyond the interpretation of the stability problem as a formal initial value problem: In weak instability, if a harmonic solution exists, the analysis holds for any initial perturbation belonging to the class of distributions Φ_m such that in the relation (7.5.2) $\psi_m = 0(1)$. Moreover, different distributions of Φ_m can be introduced an arbitrary number of times during the time history, as long as the order of magnitude requirement is respected. In weak stability, if a harmonic solution exists, the generalization is only valid until a perturbation is introduced which exceeds the critical value for instability $|a_0^{(1)}|$. As in the case of one-dimensional problems, the process of introducing different perturbations at different instants of time can be visualized as a procedure by which the constants $C_p^{(m)}$ in Eq. (7.2.2) are changed an arbitrary number of times in an arbitrary fashion during the time interval $0 \leqq t \leqq \infty$.

7.7 A simple example

As an application of the theory developed in the preceeding sections we shall consider, in a later chapter, the problem of Poiseuille-flow. In the present section we illustrate the analysis by studying a simple mathematical model in which, in contrast to the Poiseuille-flow problem, all quantities involved can be calculated explicitly without difficulty.

We consider the equation

$$\frac{I}{R}\left[\frac{\partial^2 \Phi}{\partial \eta^2} + \frac{\partial^2 \Phi}{\partial \xi^2}\right] - \frac{\partial^4 \Phi}{\partial \xi^4} - \frac{\partial \Phi}{\partial t} = \frac{\partial \Phi}{\partial \eta}\,\frac{\partial^2 \Phi}{\partial \xi^2} \tag{7.7.1}$$

with boundary conditions

$$\Phi(\xi, 0, t) = 0; \qquad \Phi(\xi, 1, t) = 1 \tag{7.7.2}$$

The basic solution satisfying Eq. (7.7.1) and the boundary conditions is given by

$$\Phi(\xi, \eta, t) = \varphi_0(\eta) = \eta \tag{7.7.3}$$

In order to study the stability of this solution we write

$$\Phi(\xi, \eta, t) = \varphi_0(\eta) + \Phi'(\xi, \eta, t) \tag{7.7.4}$$

The perturbation Φ' must then be a solution of the equation

$$\frac{1}{R}\left[\frac{\partial^2 \Phi'}{\partial \eta^2} + \frac{\partial^2 \Phi'}{\partial \xi^2}\right] - \frac{\partial^4 \Phi'}{\partial \xi^4} - \frac{\partial^2 \Phi'}{\partial \xi^2} - \frac{\partial \Phi'}{\partial t} = \frac{\partial \Phi'}{\partial \eta}\,\frac{\partial^2 \Phi'}{\partial \xi^2} \tag{7.7.5}$$

satisfying the boundary conditions

$$\Phi'(\xi, 0, t) = \Phi'(\xi, 1, t) = 0 \tag{7.7.6}$$

We consider first the linearized stability problem, that is, we neglect the right-hand-side of Eq. (7.7.5). The problem admits solution of the form

$$\Phi'(\xi, \eta, t) = e^{-ik\xi}\, e^{-\mu_p^{(k)} t}\, \varphi_p^{(k)}(\eta) \tag{7.7.7}$$

The eigenfunctions and eigenvalues are easily found:

$$\varphi_p^{(k)}(\eta) = \sin(p+1)\,\pi\,\eta \tag{7.7.8}$$

$$\mu_p^{(k)} = \frac{(p+1)^2 \pi^2 + k^2}{R} + k^2(k^2 - 1) \tag{7.7.9}$$

Clearly, all eigenfunctions are real and for every value of the parameter k there exists an infinite denumerable sequence $\mu_p^{(k)}$, with the property

$$\mu_{p+1}^{(k)} > \mu_p^{(k)}; \quad p = 0, 1, 2, \ldots \tag{7.7.10}$$

Furthermore, there exists a value R_{cr} of the parameter R such that for $R < R_{cr}$ there is stability while for $R > R_{cr}$ there is instability. More precisely, from Eq. (7.7.9) we find that $\mu_0^{(k)} < 0$ if $R > R_0$, R_0 being given by

$$R_0 = \frac{\pi^2 + k^2}{k^2(1 - k^2)} \tag{7.7.11}$$

R_{cr} is found as the minimum value of R_0. It occurs at $k = k_{cr}$,

$$(k_{cr})^2 = \pi^2\left[\sqrt{1 + \frac{1}{\pi^2}} - 1\right] \tag{7.7.12}$$

The numerical values are, with reasonable accuracy,

$$(k_{cr})^2 = 0.5; \quad R_{cr} = 40.0 \tag{7.7.13}$$

In the $k - R$ plane we have the result indicated in fig. 9.

It is seen that all hypothesis formulated in section 6.3 are verified in the present example.

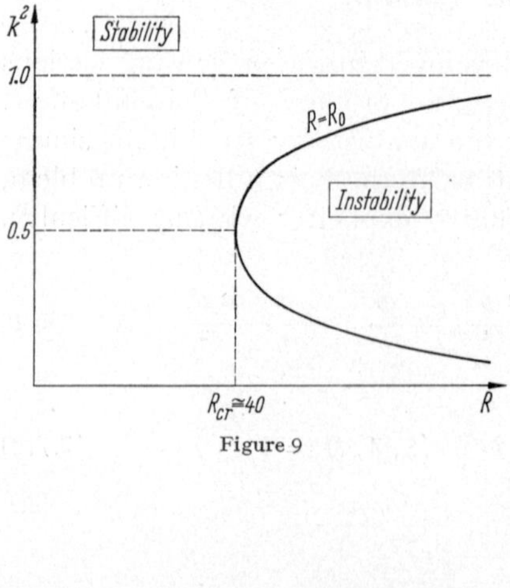

Figure 9

Turning to the non-linear problem, we study the periodic solution

$$\Phi'(\xi, \eta, t) = \sum_{-\infty}^{\infty} e^{-ik_0 m \xi} \Phi_m(\eta, t) \tag{7.7.14}$$

We choose k_0 in the vicinity of k_{cr} and R in the vicinity of R_{cr}. Substituting Eq. (7.7.14) into Eq. (7.7.5) we obtain for every component Φ_m the equation

$$\mathfrak{L}_m(\Phi_m) - \frac{\partial \Phi_m}{\partial t} = F_m \tag{7.7.15}$$

where

$$\mathfrak{L}_m = \frac{1}{R}\left[\frac{\partial^2}{\partial \eta^2} - m^2 k_0^2\right] - m^2 k_0^2 (m^2 k_0^2 - 1) \tag{7.7.16}$$

and

$$F_m = -m^2 k_0^2 \frac{\partial \Phi_0}{\partial \eta} \Phi_m - k_0^2 \sum_{n=1}^{m-1} (m-n)^2 \Phi_{m-n} \frac{\partial \Phi_n}{\partial \eta} -$$

$$- k_0^2 \sum_{n=1}^{\infty} \left[(n+m)^2 \Phi_{n+m} \frac{\partial \bar{\Phi}_n}{\partial \eta} + n^2 \bar{\Phi}_n \frac{\partial \Phi_{n+m}}{\partial \eta}\right] \tag{7.7.17}$$

We introduce the transformation

$$\Phi_m = \varepsilon\, \delta_m\, \psi_m \tag{7.7.18}$$

$$\delta_m = \varepsilon^{|m-1|}; \qquad \varepsilon^2 = |\mu_0^{(1)}| \tag{7.7.19}$$

Eq. (7.7.15) becomes

$$\mathfrak{L}_m(\psi_m) - \frac{\partial \psi_m}{\partial t} = \frac{1}{\varepsilon \delta_m} F_m \tag{7.7.20}$$

For the right-hand-side of Eq. (7.7.20) we obtain in the asymptotic approximation $\varepsilon \to 0$:

$$\frac{F_0}{\varepsilon \delta_0} = -k_0^2 \left[\psi_1 \frac{\partial \bar{\psi}_1}{\partial \eta} + \bar{\psi}_1 \frac{\partial \psi_1}{\partial \eta}\right] + 0(\varepsilon^2) \tag{7.7.21}$$

$$\frac{F_1}{\varepsilon \delta_1} = -\varepsilon^2 k_0^2 \left[\frac{\partial \psi_0}{\partial \eta} \psi_1 + 4\psi_2 \frac{\partial \bar{\psi}_1}{\partial \eta} + \bar{\psi}_1 \frac{\partial \psi_2}{\partial \eta}\right] + 0(\varepsilon^4) \tag{7.7.22}$$

$$\frac{F_2}{\varepsilon \delta_2} = -k_0^2 \psi_1 \frac{\partial \psi_1}{\partial \eta} + 0(\varepsilon^2) \tag{7.7.23}$$

etc.

Following section 7.3, we observe that

$$\psi_1(\eta, t) = \sum_{p=0}^{\infty} A_p^{(1)}(t)\, \varphi_p^{(1)}(\eta) \tag{7.7.24}$$

and

$$A_p^{(1)} = 0(\varepsilon^2) \quad \text{for} \quad p > 0 \tag{7.7.25}$$

Utilizing this result in Eqs. (7.7.21), (7.7.23) and introducing the explicit formula for the eigenfunctions (7.7.8), we find

$$\frac{1}{\varepsilon \delta_0} F_0 = |A_0^{(1)}|^2 f_0(\eta) + 0(\varepsilon^2); \qquad f_0(\eta) = -\pi k_0^2 \sin 2\pi\eta \tag{7.7.26}$$

$$\frac{1}{\varepsilon \delta_2} F_2 = [A_0^{(1)}]^2 f_2(\eta) + 0(\varepsilon^2); \qquad f_2(\eta) = -\tfrac{1}{2}\pi k_0^2 \sin 2\pi\eta \tag{7.7.27}$$

The forced solutions of Eqs. (7.7.20) now follow by the method of section 7.4; we obtain

$$\begin{aligned}\psi_0(\eta, t) &= |A_0^{(1)}|^2 g_0(\eta) + 0(\varepsilon^2)\\ g_0(\eta) &= \frac{k_0^2 \pi}{\mu_1^{(0)}} \sin 2\pi\eta\end{aligned} \tag{7.7.28}$$

$$\begin{aligned}\psi_2(\eta, t) &= [A_1^{(0)}]^2 g_2(\eta) + 0(\varepsilon^2)\\ g_2(\eta) &= \frac{k_0^2 \pi}{2\mu_1^{(2)}} \sin 2\pi\eta\end{aligned} \tag{7.7.29}$$

We turn now to the component ψ_1. Introducing Eqs. (7.7.28), (7.7.29) as well as Eqs. (7.7.24), (7.7.25) into Eq. (7.7.20), we have

$$\frac{1}{\varepsilon} F_1 = \varepsilon^2 |A_0^{(1)}|^2 A_0^{(1)} f_1(\eta) + 0(\varepsilon^4) \tag{7.7.30}$$

$$\begin{aligned}f_1(\eta) = -\pi^2 k_0^2 \Big\{ &\Big(\frac{2}{\mu_1^{(0)}} + \frac{1}{\mu_1^{(2)}}\Big) \sin\pi\eta \cos 2\pi\eta +\\ &+ \frac{2}{\mu_1^{(2)}} \sin 2\pi\eta \cos\pi\eta\Big\}\end{aligned} \tag{7.7.31}$$

Following section 7.5 we know that $A_0^{(1)}$ is a solution of the equation

$$\frac{dA_0^{(1)}}{dt} + \mu_0^{(1)} A_0^{(1)} = -\varepsilon^2 |A_0^{(1)}|^2 A_0^{(1)} \beta_0 + 0(\varepsilon^4) \tag{7.7.32}$$

where

$$\beta_0 = 2\int_0^1 f_1(\eta) \sin\pi\eta\, d\eta \tag{7.7.33}$$

Utilizing now Eq. (7.7.31) we find after a simple calculation

$$\beta_0 = \pi^2 k_0^2 \left\{\frac{1}{\mu_1^{(0)}} - \frac{1}{2\mu_1^{(2)}}\right\} \tag{7.7.34}$$

When k_0 is in the vicinity of k_{cr} and R is in the vicinity of R_{cr}, the constant β_0 has approximately the value $\beta_0 = 4$. Thus $\beta_0 > 0$ and of order unity. We conclude that there exists a harmonic solution in the case of weak instability, but since $\mu_0^{(1)}$ and β_0 are real, this solution is stationary. Consequently, in the case of weak instability

$$\lim_{t\to\infty} A_0^{(1)} = a_0^{(1)} = \mp \frac{1}{\sqrt{\beta_0}} \tag{7.7.35}$$

The asymptotic solution can now be completed by the calculation of ψ_0 and ψ_2 from Eqs. (7.7.28), (7.7.29); finally, the forced solutions for $A_p^{(1)}$, $p > 0$, as well as the components ψ_m, $m > 2$, can all be determined from the results of sections 7.2, 7.4 and 7.5.

Capter 8

STABILITY OF PERIODIC SOLUTIONS

8.1 Introduction

We have seen that under certain conditions, in the case of weak instability, the perturbation approaches in the limit as $t \to \infty$ a periodic solution of permanent type, of small and finite amplitude. However, on the basis of the analysis of Chapter 7 we can only conclude that this periodic permanent-type solution represents the limit of perturbations which are also periodic at the initial instant. On the other hand, we have already emphasized that in a proper formulation of the stability problem general *aperiodic* perturbations must be studied. Thus we should consider initial conditions such that $\Phi_k \neq 0$ for all values of k. The analysis of the preceding chapter is then simply not applicable.

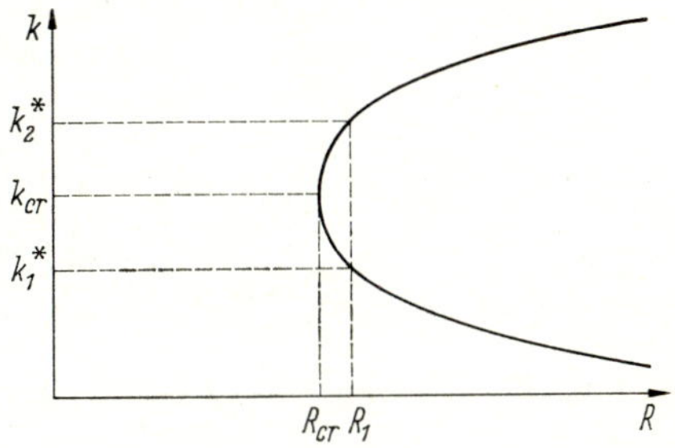

Figure 10

The periodic solutions are meaningful only if in studying the general stability problem we can demonstrate that initially *aperiodic* perturbations can approach in the limit $t \to \infty$ a periodic solution. More precisely, let us consider in the $k - R$ plane the case $R = R_1, R_1 > R_{cr}$, $|R_1 - R_{cr}| \ll 1$, as indicated in fig. 10.

We assume that $Re(\beta_0) > 0$, so that for a fixed value $R = R_1$ we have the possibility od periodic solutions with period $2\pi/k_0$ for all values of k_0 satisfying the relation $k_1^* < k_0 < k_2^*$.

If a periodic solution is to be the limit for $t \to \infty$ of some initially aperiodic perturbations, we must have

$$\lim_{t \to \infty} \Phi_k = 0 \quad \text{if} \quad k \neq m\, k_0; \qquad m = 0, 1, 2 \ldots \tag{8.1.1}$$

In other words, the periodic solution with period $2\pi/k_0$ must be stable with respect to all perturbations Φ_k, $k \neq m\, k_0$. This stability problem is the subject of the present chapter. The necessary analysis proves to be of considerable complexity, but it ultimately leads to a simple remarkable result, which can be summarized as follows:

We define a suitably small level of aperiodic perturbations Φ_k, $k \neq m\, k_0$, and show that: *In the interval $k_1^* \leq k \leq k_2^*$ there exists a subinterval of values of k such that if k_0 is in that subinterval, the corresponding periodic solution is stable with respect to the aperiodic perturbations.* We also determine the exact extent that stable subinterval.

It is in the light of this result that an interpretation can be attached to the periodic solutions of chapter 7: those among them which are stable in the sense defined above represent the limit for $t \to \infty$ of initially aperiodic perturbations and can be interpreted as solutions of the general problem of stability.

This does not mean, of course, that the stable periodic solutions represent the limit for $t \to \infty$ of any arbitrary aperiodic initial perturbations; it does mean however that the stable periodic solutions are not 'isolated', but represent the limit of some aperiodic initial perturbations.

In what follows the stability problem described here is formulated and analysed; in a final section we summarize and discuss the results.

8.2 Formulation of the stability problem

In the case of general aperiodic perturbations none of the Fourier-components Φ_k vanishes and we must consider the complete system of equations (6.4.6), (6.4.7), (6.4.8), allowing the parameter k to take all real values. We introduce the transformation

$$\Phi_{m k_0} = \varepsilon\, \delta_m\, \psi_m; \qquad \delta_m = \varepsilon^{|m-1|}; \qquad m = 0, 1, 2 \ldots \tag{8.2.1}$$

$$\Phi_k = \varepsilon^3\, \psi_k \quad \text{if} \quad k \neq m\, k_0 \tag{8.2.2}$$

$$\varepsilon^2 = |\tau_0^{(k_0)}|; \qquad \tau_0^{(k_0)} = Re[\mu_0^{(k_0)}] \tag{8.2.3}$$

which is the transformation already utilized in the theory of periodic solutions. If we assume that ψ_m and ψ_k are of order unity (at least initially) then we study the stability of the periodic solution with respect to perturbations Φ_k, $k \neq m\, k_0$, of which the level is chosen to be of order ε^3.

The choice of the level of perturbations may seem quite arbitrary at this stage but in fact it is not. Its significance will be clarified shortly; for the moment we shall proceed formally with the transformation defined above, introduce it into Eqs. (6.4.6), (6.4.7), (6.4.8) and consider the asymptotic approximation of the equations for $\varepsilon \to 0$.

We start with the values $k = m\, k_0$. Performing the operations described above we find

$$\left[\mathfrak{L}_0 - \frac{\partial}{\partial t} S_0\right] \psi_0 = \sum_{l=1}^{N} \left\{\bar{P}_{k_0}^{(l)}(\bar{\psi}_1)\, Q_{k_0}^{(l)}(\psi_1) + \bar{Q}_{k_0}^{(l)}(\bar{\psi}_1)\, P_{k_0}^{(l)}(\psi_1)\right\} + 0(\varepsilon^2), \tag{8.2.4}$$

$$\left[\mathfrak{L}_{k_0} - \frac{\partial}{\partial t} S_{k_0}\right] \psi_1 = \varepsilon^2 \sum_{l=1}^{N} \left\{P_0^{(l)}(\psi_0)\, Q_{k_0}^{(l)}(\psi_1) + Q_0^{(l)}(\psi_0)\, P_{k_0}^{(l)}(\psi_1) + \right.$$
$$\left. + \bar{P}_{k_0}^{(l)}(\bar{\psi}_1)\, Q_{2k_0}^{(l)}(\psi_2) + \bar{Q}_{k_0}^{(l)}(\bar{\psi}_1)\, P_{2k_0}^{(l)}(\psi_2)\right\} + 0(\varepsilon^4) \tag{8.2.5}$$

$$\left[\mathfrak{L}_{2k_0} - \frac{\partial}{\partial t} S_{2k_0}\right] \psi_2 = \sum_{l=0}^{N} P_{k_0}^{(l)}(\psi_1)\, Q_{k_0}^{(l)}(\psi_1) + 0(\varepsilon^2) \tag{8.2.6}$$

etc.

With a slight change in notation resulting from the greater complexity of the present problem, these are the equations of periodic solutions of Chapter 7. We see that when the level of perturbations Φ_k, $k \neq m\,k_0$, is limited by Eq. (8.2.2), the components of the periodic solution $\Phi_{m k_0}$ are not directly affected by the presence of perturbations. Solutions of Eqs. (8.2.4), (8.2.5), (8.2.6) have been obtained in Chapter 7; we shall summarize them here for future convenience in the notation that will be used throughout this chapter. For $\varepsilon \to 0$ we have:

$$\psi_0(\eta, t) = |A_0^{(k_0)}(t)|^2\, g_0(\eta) + 0(\varepsilon^2) \tag{8.2.7}$$

$$\psi_1(\eta, t) = A_0^{(k_0)}(t)\, \varphi_0^{(k_0)}(\eta) + 0(\varepsilon^2) \tag{8.2.8}$$

$$\psi_2(\eta, t) = [A_0^{(k_0)}(t)]^2\, g_2(\eta) + 0(\varepsilon^2) \tag{8.2.9}$$

The amplitude function $A_0^{(k_0)}$ is a solution of the equation

$$\frac{d A_0^{(k_0)}}{dt} + \mu_0^{(k_0)} A_0^{(k_0)} = -\varepsilon^2 \beta_0 A_0^{(k_0)} |A_0^{(k_0)}|^2 + 0(\varepsilon^4) \tag{8.2.10}$$

where

$$\beta_0 = \sum_{l=1}^{N} \int_0^1 f_1^{(l)}(\eta)\, \bar{\varphi}_0^{(k_0)}(\eta)\, d\eta \tag{8.2.11}$$

$$f_1^{(l)} = P_0^{(l)}(g_0)\, Q_{k_0}^{(l)}(\varphi_0^{(k_0)}) + Q_0^{(l)}(g_0)\, P_{k_0}^{(l)}(\varphi_0^{(k_0)}) + P_{2k_0}^{(l)}(g_2)\, \bar{Q}_{k_0}^{(l)}(\bar{\varphi}_0^{(k_0)}) + Q_{2k_0}^{(l)}(g_2)\, \bar{P}_{k_0}^{(l)}(\bar{\varphi}_0^{(k_0)}) \tag{8.2.12}$$

The functions g_0 and g_2 are defined by the eigenfunction expansions

$$g_n(\eta) = -\sum_{p=0}^{\infty} \left[\int_0^1 \sum_{l=1}^{N} f_n^{(l)}(\eta')\, \bar{\varphi}_p^{(n k_0)}(\eta')\, d\eta' \right] \frac{\varphi_p^{(n k_0)}(\eta)}{\mu_p^{(n k_0)} - i n \nu_0^{(k_0)}} \tag{8.2.13}$$

where

$$\mathfrak{f}_0^{(l)}(\eta) = \bar{P}_{k_0}^{(l)}(\bar{\varphi}_0^{(k_0)})\, Q_{k_0}^{(l)}(\varphi_0^{(k_0)}) + \bar{Q}_{k_0}^{(l)}(\bar{\varphi}_0^{(k_0)})\, P_{k_0}^{(l)}(\varphi_0^{(k_0)}) \tag{8.2.14}$$

$$\mathfrak{f}_2^{(l)}(\eta) = P_{k_0}^{(l)}(\varphi_0^{(k_0)})\, Q_{k_0}^{(l)}(\varphi_0^{(k_0)}) \tag{8.2.15}$$

The behaviour of the components $\Phi_{m k_0}$ can now be considered completely determined and we turn to the study of perturbations Φ_k, $k \neq m\,k_0$. Introducing the transformation (8.2.4), (8.2.5), (8.2.6) we obtain the following asymptotic development from Eqs. (6.4.6), (6.4.8).

$$\begin{aligned}\left[\mathfrak{L}_k - \frac{\partial}{\partial t} S_k\right] \psi_k = {} & \varepsilon \sum_{l=1}^{N} \{P_{k_0}^{(l)}(\psi_1)\, Q_{k-k_0}^{(l)}(\psi_{k-k_0}) + Q_{k_0}^{(l)}(\psi_1)\, P_{k-k_0}^{(l)}(\psi_{k-k_0}) + \\ & + \bar{P}_{k_0}^{(l)}(\bar{\psi}_1)\, Q_{k+k_0}^{(l)}(\psi_{k+k_0}) + \bar{Q}_{k_0}^{(l)}(\bar{\psi}_1)\, P_{k+k_0}^{(l)}(\psi_{k+k_0})\} + \\ & + \varepsilon^2 \sum_{l=1}^{N} \{P_{2k_0}^{(l)}(\psi_2)\, \bar{Q}_{2k_0-k}^{(l)}(\bar{\psi}_{2k_0-k}) + Q_{2k_0}^{(l)}(\psi_2)\, \bar{P}_{2k_0-k}^{(l)}(\bar{\psi}_{2k_0-k}) + \\ & + \bar{P}_{2k_0}^{(l)}(\bar{\psi}_2)\, Q_{2k_0+k}^{(l)}(\psi_{2k_0+k}) + \bar{Q}_{2k_0}^{(l)}(\bar{\psi}_2)\, P_{2k_0+k}^{(l)}(\psi_{2k_0+k}) + \\ & + P_0^{(l)}(\psi_0)\, Q_k^{(l)}(\psi_k) + Q_0^{(l)}(\psi_0)\, P_k^{(l)}(\psi_k)\} + 0(\varepsilon^3)\end{aligned} \tag{8.2.16}$$

In view of the complexity of Eq. (8.2.16) it is not surprising that the study of the stability of periodic solutions requires a lengthy delicate analysis. Nevertheless, we shall see that ultimately the problem can be reduced to a rather simple form.

Analysing the right-hand-side of Eq. (8.2.16) we note that the perturbation component Φ_k, $k \neq m\,k_0$, interacts with the perturbation components $\Phi_{k_0 \mp k}$ and $\Phi_{2k_0 \mp k}$. However, and this is essential, Eq. (8.2.16) is linear in the perturbation quantities, since the functions Φ_0, Φ_1 and Φ_2 have already been determined.

Now Eq. (8.2.16) is valid for perturbations limited (in order of magnitude) by Eq. (8.2.2). If we had started the analysis with perturbations of a higher level, the asymptotic expansion would have yielded non-linear equations for the perturbation quantities. Eq. (8.2.2) should thus be considered as a condition needed to assure that the asymptotic development in the stability problem will lead to its linearization.

Usually, when stability problems are studied in linear approximation, the linearization procedure is justified by the statement that perturbations are assumed to be 'infinitesimally small'. In the present problem we could have started from that statement and linearize the equations for Φ_k, $k \neq m\,k_0$, formally. This would have led to a result equivalent with Eq. (8.2.16), but the procedure is not satisfactory, for the following reasons:

We study the stability of a solution which itself has a small amplitude $(\Phi_{m k_0} = O(\varepsilon))$. Now perturbations can be chosen of smaller order of magnitude, independent of ε, and if that order of magnitude is chosen 'arbitrarily small' we have infinitesimal perturbations and formal linearization of the stability problem. But it is desirable to establish *how small* the perturbations must be chosen in comparison with ε for the linearization to be a valid procedure. This is exactly what has been accomplished by the transformation (8.2.2) and the subsequent asymptotic expansion.

We note that although the problem is now linear, we still have to solve an *infinite* system of equations (8.2.16). Furthermore, we must analyse the stability of all the periodic solutions corresponding to the values of k_0 contained in the interval $k_1^* < k_0 < k_2^*$. It proves convenient to develop the method of solution while considering a particular case. Below, unless we state otherwise, we study the case $k_0 = k_{cr}$. We shall find ultimately that the results obtained are easily extended to other values of k_0.

Before undertaking the study of Eqs. (8.2.16), however, it is necessary to analyse the small parameters of the problem in some detail. This is done in the next section.

8.3 Analysis of small parameters

Generalizing the notation introduced earlier we write

$$\tau_n^{(k)} = Re[\mu_n^{(k)}]; \qquad \nu_n^{(k)} = Im(\mu_n^{(k)}) \tag{8.3.1}$$

The eigenvalues of the linearized problem are ordered in such a way that for every fixed value of k

$$\tau_{n+1}^{(k)} > \tau_n^{(k)}; \qquad n = 0, 1, 2, \ldots \tag{8.3.2}$$

If now, in fig. 10, R is chosen sufficiently close to R_{cr}, we have

$$\tau_n^{(k)} > 0 \quad \text{and} \quad \tau_n^{(k)} = 0(1) \tag{8.3.3}$$

for all values of k and all $n > 0$. We study now, in some detail, the behaviour of the parameter $\tau_0^{(k)}$.

We choose $R > R_{cr}$ and consider $\tau_0^{(k)}$ as a function of k for a fixed value of R. Then we must have a curve as sketched in fig. 11.

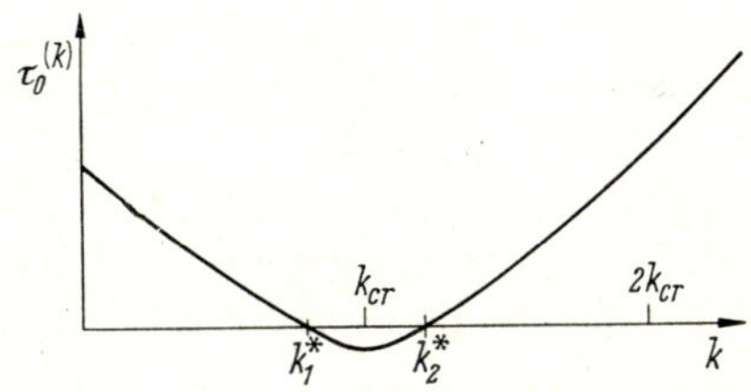

Figure 11

$\tau_0^{(k)}$ is negative for $k_1^* < k < k_2^*$, and the minimum value reached in this interval is a small quantity.

For $k > k_2^*$ and $k < k_1^*$, $\tau_0^{(k)}$ is positive and in the neighbourhood of $k = 0$ and $k = 2k_{cr}$, $\tau_0^{(k)}$ reaches values of order unity, compared to the minimum value in the interval $k_1^* < k < k_2^*$. For the purpose of further analysis, however, these qualitative statements are not sufficient and a formula is needed to define more precisely the behaviour of $\tau_0^{(k)}$ with k.

As a preliminary, consider $\tau_0^{(k)}$ as a function of R, for a fixed value of $k = k_{cr}$. Assuming that in the neighbourhood of $k = k_{cr}$ and $R = R_{cr}$, $\tau_0^{(k)}$ is an analytic function of k and R, we can write the Taylor's series

$$\tau_0^{(k_{cr})} = -\alpha_1 (R - R_{cr}) + \cdots \tag{8.3.4}$$

where

$$\alpha_1 = \left[\frac{\partial \tau_0^{(k_{cr})}}{\partial R}\right]_{R = R_{cr}} \tag{8.3.5}$$

Obviously, the constant α_1 is positive. In accordance with Eq. (8.2.3) we now conclude

$$|R - R_{cr}| = 0(\varepsilon^2) \tag{8.3.6}$$

In this way the quantity $R - R_{cr}$ is related to the small parameter ε introduced earlier, and the asymptotic expansion $\varepsilon \to 0$ can be interpreted as an expansion for $R \to R_{cr}$.

The result (8.3.6) allows us to evaluate the extent of the interval $k_1^* < k < k_2^*$ in fig. 11. For that purpose we consider the critical curve in fig. 10. If that curve is sufficiently regular in the neighbourhood of $k = k_{cr}$, then it can be represented by the Taylor's series

$$R - R_{cr} = \gamma_2 (k - k_{cr})^2 + \gamma_3 (k - k_{cr})^3 + \cdots \tag{8.3.7}$$

where γ_2, γ_3, etc are constants and $\gamma_2 > 0$. In fact, if $R = R_0(k) = 0$ is the equation of the critical curve, then

$$\gamma_2 = \frac{1}{2}\left(\frac{d^2 R_0}{d k^2}\right)_{k=k_{cr}} \tag{8.3.8}$$

It follows from Eq. (8.3.7) that for values of R satisfying Eq. (8.3.6) the two points of the critical curve are given by

$$k^* - k_{cr} = \mp \frac{1}{\sqrt{\gamma_2}} \sqrt{R - R_{cr}} + 0(\varepsilon^2) \tag{8.3.9}$$

Consequently, the interval of negative values of $\tau_0^{(k)}$ in fig. 11 is defined by

$$\begin{aligned} k_1^* - k_{cr} &= - \frac{1}{\sqrt{\gamma_2}} \sqrt{R - R_{cr}} + 0(\varepsilon^2) \\ k_2^* - k_{cr} &= + \frac{1}{\sqrt{\gamma_2}} \sqrt{R - R_{cr}} + 0(\varepsilon^2) \end{aligned} \tag{8.3.10}$$

It is seen that the length of that interval is of order of magnitude ε.

We now seek a formula for $\tau_0^{(k)}$ valid in the domain $k - k_{cr} = 0(\varepsilon)$. For this purpose we introduce the transformation

$$k = k_{cr} + \varepsilon\,\sigma; \qquad \varepsilon = \sqrt{|\tau_0^{(k_{cr})}|} \tag{8.3.11}$$

The Taylor's series for $\tau_0^{(k)}$, in the vicinity of $k = k_{cr}$, reads in this notation

$$\tau_0^{(k)} = \tau_0^{(k_{cr})} \{1 + \varepsilon\, a_1 \sigma + \varepsilon^2 a_2 \sigma^2 + \varepsilon^3 a_3 \sigma^3 + \cdots\} \tag{8.3.12}$$

The coefficients a_1, a_2, etc. are established by a simple argument, supposing sufficient regularity of the curve $\tau_0^{(k)}$. If in fact we assume that

$$\lim_{\varepsilon \to 0} \frac{\partial^n \tau_0^{(k)}}{\partial k^n}$$

is finite, we can immediately conclude that $a_n = 0(\varepsilon^{n-2})$ for $n \geqq 2$.

On the other hand, by requiring that $\tau_0^{(k)} = 0$ for $k = k_1^*$ and $k = k_2^*$ we find that a_1 is at most of order of magnitude unity, while a_2 is given by

$$a_2 = - \frac{\alpha_1 \gamma_2}{\varepsilon^2} + 0(1) \tag{8.3.13}$$

Thus, in asymptotic approximation we have

$$\tau_0^{(k)}(\sigma) = \tau_0^{(k_{cr})} \{1 - (\alpha_1 \gamma_2)\, \sigma^2 + 0(\varepsilon)\} \tag{8.3.14}$$

The result (8.3.14) can be rewritten in the form

$$\tau_0^{(k)} = \tau_0^{(k_{cr})} \left\{ 1 - \left(\frac{k - k_{cr}}{k^* - k_{cr}} \right)^2 + 0(\varepsilon) \right\} \tag{8.3.15}$$

to show more clearly that in fig. 11 the curve $\tau_0^{(k)}$ has been approximated in the neighbourhood of $k = k_{cr}$ by a parabola of which the two zeros are given by Eq. (8.3.10).

The parabolic approximation in fig. 11 is quite natural and could have been proposed at the outset on intuitive grounds. In the context of an asymptotic theory, however, it is desirable to know the order of magnitude of the terms that have been neglected when an approximation of the type (8.3.15) is used. That is why the parabolic approximation was deduced here from the asymptotic expansion for $\varepsilon \to 0$, showing that in the asymptotic theory, the approximation (8.3.14) is indeed the proper representation of $\tau_0^{(k)}$ in the region $k - k_{cr} = 0(\varepsilon)$.

8.4 Perturbations in the region $\tau_0^{(k)} = 0(1)$

We now consider the perturbations ψ_k corresponding to values of k such that $\tau_0^{(k)} = 0(1)$ as $\varepsilon \to 0$. Abbreviating, we write Eq. (8.2.16) as follows

$$\left[\mathfrak{L}_k - \frac{\partial}{\partial t} S_k \right] \psi_k = \varepsilon\, F_k [\psi_0, \psi_1, \psi_2; \psi_{k_0 \mp k}, \varepsilon\, \psi_k, \varepsilon\, \psi_{2k_0 \pm k}] \tag{8.4.1}$$

We introduce the eigenfunction expansion

$$\psi_k(\eta, t) = \sum_{n=0}^{\infty} A_n^{(k)}(t)\, \varphi_n^{(k)}(\eta) \tag{8.4.2}$$

which leads to

$$\frac{d A_n^{(k)}}{dt} + \mu_n^{(k)} A_n^{(k)} = -\varepsilon \int_0^1 F_k \tilde{\varphi}_n^{(k)}\, d\eta \tag{8.4.3}$$

If now $\tau_0^{(k)} = 0(1)$ and is positive, the solutions of Eq. (8.4.3) must be of the form

$$A_n^{(k)} = C_n^{(k)} e^{-\mu_n^{(k)} t} + 0(\varepsilon) \tag{8.4.4}$$

where $C_n^{(k)}$ are constants to be determined from initial conditions. Clearly, in the first approximation, the perturbation tends to zero as $t \to \infty$.

In order to evaluate the term of order ε in Eq. (8.4.4) we need to retain in the definition of F_k only the terms of order of magnitude unity. Comparing with Eq. (8.2.16) we have

$$F_k = \sum_{l=1}^{N} \{P_{k_0}^{(l)}(\psi_1)\, Q_{k-k_0}^{(l)}(\psi_{k-k_0}) + Q_{k_0}^{(l)}(\psi_1)\, P_{k-k_0}^{(l)}(\psi_{k-k_0}) +$$

$$+ \bar{P}_{k_0}^{(l)}(\bar{\psi}_1)\, Q_{k+k_0}^{(l)}(\psi_{k+k_0}) + \bar{Q}_{k_0}^{(l)}(\bar{\psi}_1)\, P_{k+k_0}^{(l)}(\psi_{k+k_0})\} + 0(\varepsilon) \qquad (8.4.5)$$

Thus, the term of order ε in Eq. (8.4.4) is of the nature of a forced solution, depending on ψ_{k+k_0} and ψ_{k-k_0}. We develop the explicit formulas for these forced solutions later but to do so we generally need to know the solutions for ψ_k in regions where $\tau_0^{(k)} \neq 0(1)$. If, for example, we consider $k = 2k_0 + \varepsilon\,\sigma$, the forced solution in Eq. (8.4.4) will depend on $\psi_{3k+\varepsilon\sigma}$ and $\psi_{k_0+\varepsilon\sigma}$, while $\tau_0^{(k_0+\varepsilon\sigma)} = 0(\varepsilon^2)$.

In fact, and this is not too surprising, the study of perturbations corresponding to k values such that $\tau_0^{(k)} = 0(\varepsilon^2)$ constitutes the essential part of the stability analysis of periodic solutions.

8.5 Perturbations in the region $\tau_0^{(k)} = 0(\varepsilon^2)$

We consider k values in the domain $k - k_{cr} = 0(\varepsilon)$. It is then convenient to write

$$k_{\prime} = k_0 + \varepsilon\,\sigma$$
$$k_{\prime\prime} = k_0 - \varepsilon\,\sigma \qquad (8.5.1)$$

Taking for k the value $k_{\prime}$ we have from Eq. (8.2.16), in the notation (8.5.1), the equation

$$\left[\mathfrak{L}_{k_\prime} - \frac{\partial}{\partial t} S_{k_\prime}\right] \psi_{k_\prime} = \varepsilon^2 \sum_{l=1}^{N} F_{k_\prime}^{(l)} \qquad (8.5.2)$$

$$F_{k_\prime}^{(l)} = \frac{1}{\varepsilon} \{P_{k_0}^{(l)}(\psi_1)\, Q_{\varepsilon\sigma}^{(l)}(\psi_{\varepsilon\sigma}) + Q_{k_0}^{(l)}(\psi_1)\, P_{\varepsilon\sigma}^{(l)}(\psi_{\varepsilon\sigma}) +$$

$$+ \bar{P}_{k_0}^{(l)}(\bar{\psi}_1)\, Q_{2k_0+\varepsilon\sigma}^{(l)}(\psi_{2k_0+\varepsilon\sigma}) + \bar{Q}_{k_0}^{(l)}(\bar{\psi}_1)\, P_{2k_0+\varepsilon\sigma}^{(l)}(\psi_{2k_0+\varepsilon\sigma})\} +$$

$$+ \{P_0^{(l)}(\psi_0)\, Q_{k_\prime}^{(l)}(\psi_{k_\prime}) + Q_0^{(l)}(\psi_0)\, P_{k_\prime}^{(l)}(\psi_{k_\prime}) +$$

$$+ P_{2k_0}^{(l)}(\psi_2)\, \bar{Q}_{k_{\prime\prime}}^{(l)}(\bar{\psi}_{k_{\prime\prime}}) + Q_{2k_0}^{(l)}(\psi_2)\, \bar{P}_{k_{\prime\prime}}^{(l)}(\bar{\psi}_{k_{\prime\prime}}) +$$

$$+ \bar{P}_{2k_0}^{(l)}(\bar{\psi}_2)\, Q_{3k_0+\varepsilon\sigma}^{(l)}(\psi_{3k_0+\varepsilon\sigma}) + \bar{Q}_{2k_0}^{(l)}(\bar{\psi}_2)\, P_{3k_0+\varepsilon\sigma}^{(l)}(\psi_{3k_0+\varepsilon\sigma})\} + 0(\varepsilon) \qquad (8.5.3)$$

We see that the perturbation $\psi_{k_\prime}$, for which $\tau_0^{(k_\prime)} = 0(\varepsilon^2)$, interacts with another perturbation $\psi_{k_{\prime\prime}}$ which belongs to the region $\tau_0^{(k)} = 0(\varepsilon^2)$. Furthermore, $\psi_{k_\prime}$ interacts with perturbations $\psi_{\varepsilon\sigma}$, $\psi_{2k_0+\varepsilon\sigma}$ and $\psi_{3k_0+\varepsilon\sigma}$ which belong to the region of k values such that $\tau_0^{(k)} = 0(1)$.

The equation for $\psi_{k_{\prime\prime}}$ is obtained from (8.5.2), (8.5.3) when $\varepsilon\,\sigma$ is replaced by $-\varepsilon\,\sigma$. It is easily seen that $\psi_{k_{\prime\prime}}$ interacts with $\psi_{k_\prime}$, and also with $\psi_{\varepsilon\sigma}$, $\psi_{2k_0-\varepsilon\sigma}$ and $\psi_{3k_0-\varepsilon\sigma}$.

The equations for other perturbation components involved in the analysis of $\psi_{k_\prime}$ and $\psi_{k_{\prime\prime}}$ follow again from Eq. (8.2.16), and are given by

$$\left[\mathfrak{L}_{\varepsilon\sigma} - \frac{\partial}{\partial t} S_{\varepsilon\sigma}\right] \psi_{\varepsilon\sigma} = \varepsilon \sum_{l=1}^{N} \{P_{k_0}^{(l)}(\psi_1)\, \bar{Q}_{k_{\prime\prime}}^{(l)}(\bar{\psi}_{k_{\prime\prime}}) + Q_{k_0}^{(l)}(\psi_1)\, \bar{P}_{k_{\prime\prime}}^{(l)}(\bar{\psi}_{k_{\prime\prime}}) + $$

$$+ \bar{P}_{k_0}^{(l)}(\bar{\psi}_1)\, Q_{k_\prime}^{(l)}(\psi_{k_\prime}) + \bar{Q}_{k_0}^{(l)}(\bar{\psi}_1)\, P_{k_\prime}^{(l)}(\psi_{k_\prime})\} + 0(\varepsilon^2) \tag{8.5.4}$$

$$\left[\mathfrak{L}_{2k_0+\varepsilon\sigma} - \frac{\partial}{\partial t} S_{2k_0+\varepsilon\sigma}\right] \psi_{2k_0+\varepsilon\sigma} = \varepsilon \sum_{l=1}^{N} \{P_{k_0}^{(l)}(\psi_1)\, Q_{k_\prime}^{(l)}(\psi_{k_\prime}) + $$

$$+ Q_{k_0}^{(l)}(\psi_1)\, P_{k_\prime}^{(l)}(\psi_{k_\prime}) + \bar{P}_{k_0}^{(l)}(\bar{\psi}_1)\, Q_{3k_0+\varepsilon\sigma}^{(l)}(\psi_{3k_0+\varepsilon\sigma}) + $$

$$+ \bar{Q}_{k_0}^{(l)}(\bar{\psi}_1)\, P_{3k_0+\varepsilon\sigma}^{(l)}(\psi_{3k_0+\varepsilon\sigma})\} + 0(\varepsilon^2) \tag{8.5.5}$$

$$\left[\mathfrak{L}_{3k_0+\varepsilon\sigma} - \frac{\partial}{\partial t} S_{3k_0+\varepsilon\sigma}\right] \psi_{3k_0+\varepsilon\sigma} = \varepsilon \sum_{l=1}^{N} \{P_{k_0}^{(l)}(\psi_1)\, Q_{2k_0+\varepsilon\sigma}^{(l)}(\psi_{2k_0+\varepsilon\sigma}) + $$

$$+ Q_{k_0}^{(l)}(\psi_1)\, P_{2k_0+\varepsilon\sigma}^{(l)}(\psi_{2k_0+\varepsilon\sigma}) + \bar{P}_{k_0}^{(l)}(\bar{\psi}_1)\, Q_{4k_0+\varepsilon\sigma}^{(l)}(\psi_{4k_0+\varepsilon\sigma}) + $$

$$+ \bar{Q}_{k_0}^{(l)}(\bar{\psi}_1)\, P_{4k_0+\varepsilon\sigma}^{(l)}(\psi_{4k_0+\varepsilon\sigma})\} + 0(\varepsilon^2) \tag{8.5.6}$$

Finally, the equations for $\psi_{2k_0-\varepsilon\sigma}$ and $\psi_{3k_0-\varepsilon\sigma}$ follow from Eqs. (8.5.5), (8.5.6) when $\varepsilon\sigma$ is replaced by $-\varepsilon\sigma$.

It should be noted that at this stage we still have an infinite system of equations, since the equation for $\psi_{3k_0\mp\varepsilon\sigma}$ still contains $\psi_{4k_0\mp\varepsilon\sigma}$, and continuing we find $\psi_{4k_0\mp\varepsilon\sigma}$ depending on $\psi_{5k_0\mp\varepsilon\sigma}$, etc.

8.6 Reduction of the system of equations

In section 8.4 we have found that if $\tau_0^{(k)} = 0(1)$ the solution for ψ_k is composed of an infinite series of terms decaying exponentially with time, and of a forced solution of order of magnitude ε. Applying this result to $\psi_{\varepsilon\sigma}$, $\psi_{2k_0\mp\varepsilon\sigma}$, $\psi_{3k_0\mp\varepsilon\sigma}$ we write

$$\psi_k^{(l)} = \sum_{n=0}^{\infty} C_n^{(k)}\, e^{-\mu_n^{(k)} t}\, \varphi_n^{(k)}(\eta) \tag{8.6.1}$$

The solutions of Eqs. (8.5.4) to (8.5.6) can then symbolically be written in the form

$$\left.\begin{aligned} \psi_{\varepsilon\sigma} &= \psi_{\varepsilon\sigma}^{(l)} + \varepsilon\, \psi_{\varepsilon\sigma}^{(f)}[\psi_{k_\prime};\, \psi_{k_{\prime\prime}}] \\ \psi_{2k_0+\varepsilon\sigma} &= \psi_{2k_0+\varepsilon\sigma}^{(l)} + \varepsilon\, \psi_{2k_0+\varepsilon\sigma}^{(f)}[\psi_{k_\prime};\, \psi_{3k_0+\varepsilon\sigma}] \\ \psi_{2k_0-\varepsilon\sigma} &= \psi_{2k_0-\varepsilon\sigma}^{(l)} + \varepsilon\, \psi_{2k_0-\varepsilon\sigma}^{(f)}[\psi_{k_{\prime\prime}};\, \psi_{3k_0-\varepsilon\sigma}] \\ \psi_{3k_0\mp\varepsilon\sigma} &= \psi_{3k_0\mp\varepsilon\sigma}^{(l)} + 0(\varepsilon) \end{aligned}\right\} \tag{8.6.2}$$

where $\psi_k^{(f)}$ represent the forced solutions while $\psi_k^{(l)}$ are fully determined from the initial conditions.

It is clear now that in calculating the forced solutions for $\psi_{2k_0\mp\varepsilon\sigma}$ in the asymptotic approximation we can neglect the forced solution for

$\psi_{3k_0 \mp \varepsilon\sigma}$ which is of order of magintude ε. The forced solutions for $\psi_{2k_0 \mp \varepsilon\sigma}$ appear thus as functions of $\psi_{k_,}$ and $\psi_{k_{,,}}$ only.

Similarly, in Eq. (8.5.3) the forced solutions for $\psi_{3k_0 \mp \varepsilon\sigma}$ can again be neglected, so that, as far as the forced solutions are concerned, $\psi_{k_,}$ and $\psi_{k_{,,}}$ depends only on $\psi_{\varepsilon\sigma}$ and $\psi_{2k_0 \mp \varepsilon\sigma}$.

In this way, the infinite system of equations reduces in the asymptotic theory to a system of *five* equations for the five unknowns $\psi_{k_,}$, $\psi_{k_{,,}}$, $\psi_{\varepsilon\sigma}$ and $\psi_{2k_0 \mp \varepsilon\sigma}$. But before actually performing the calculations it is useful to study in the some detail the role of the exponential series of the type (8.6.1).

Suppose that we have determined the explicit form of the forced solutions $\psi^{(f)}_{\varepsilon\sigma}$, $\psi^{(f)}_{2k_0 \mp \varepsilon\sigma}$ as functions of $\psi_{k_,}$ and $\psi_{k_{,,}}$ and introduced the solutions (8.6.2) into Eqs. (8.5.2), (8.5.3). As a result of those operations we obtain two equations for $\psi_{k_,}$ and $\psi_{k_{,,}}$, which symbolically can be written as follows

$$\left[\mathfrak{L}_{k_,} - \frac{\partial}{\partial t} S_{k_,}\right] \psi_{k_,} = T^{(1)}_{k_,}(\psi_{k_,}) + T^{(2)}_{k_,}(\psi_{k_{,,}}) + U_{k_,}[\psi^{(l)}_{\varepsilon\sigma};\ \psi^{(l)}_{2k_0+\varepsilon\sigma};\ \psi^{(l)}_{3k_0+\varepsilon\sigma}] \tag{8.6.3}$$

$$\left[\mathfrak{L}_{k_{,,}} - \frac{\partial}{\partial t} S_{k_{,,}}\right] \psi_{k_{,,}} = T^{(1)}_{k_{,,}}(\psi_{k_{,,}}) + T^{(2)}_{k_{,,}}(\psi_{k_,}) + U_{k_{,,}}[\psi^{(l)}_{\varepsilon\sigma};\ \psi^{(l)}_{2k_0-\varepsilon\sigma};\ \psi^{(l)}_{3k_0-\varepsilon\sigma}] \tag{8.6.4}$$

where $T^{(1)}_{k_,}(\psi_{k_,})$ is a linear expression in $\psi_{k_,}$, $T^{(2)}_{k_,}(\psi_{k_{,,}})$ a linear expression in $\psi_{k_{,,}}$, etc. these expressions arising from the substitution of the forced solutions $\psi^{(f)}_{\varepsilon\sigma}$, $\psi^{(f)}_{2k_0 \mp \varepsilon\sigma}$ into Eq. (8.5.3). On the other hand, the expressions $U_{k_,}$ and $U_{k_{,,}}$ consist of infinite series ot the type (8.6.1) of which all coefficients can be considered to be known.

In other words, in Eqs. (8.6.3), (8.6.4) we obtain a system of two coupled non-homogeneous linear equations for $\psi_{k_,}$ and $\psi_{k_{,,}}$. The solution of the problem for $\psi_{k_,}$, $\psi_{k_{,,}}$ consists of the solution of the homogeneous problem and a particular solution, function of $U_{k_,}$ and $U_{k_{,,}}$. However, we know that $U_{k_,} \to 0$ and $U_{k_{,,}} \to 0$ as $t \to \infty$. Under those conditions the behaviour of $\psi_{k_,}$ and $\psi_{k_{,,}}$ for $t \to \infty$ is determined by the solution of the homogeneous problem. It follows that in performing the analysis we only need to substitute in Eqs. (8.5.2), (8.5.3) the forced solutions for $\psi_{\varepsilon\sigma}$ and $\psi_{2k_0 \mp \varepsilon\sigma}$.

Returning to Eqs. (8.5.2), (8.5.3) we introduce the eigenfunction expansions

$$\begin{aligned} \psi_{k_,}(\eta, t) &= \sum_{n=0}^{\infty} A^{(k_,)}_n(t)\, \varphi^{(k_,)}_n(\eta) \\ \psi_{k_{,,}}(\eta, t) &= \sum_{n=0}^{\infty} A^{(k_{,,})}_n(t)\, \varphi^{(k_{,,})}_n(\eta) \end{aligned} \tag{8.6.5}$$

we then obtain

$$\frac{dA_n^{(k_,)}}{dt} + \mu_n^{(k_,)} A_n^{(k_,)} = -\varepsilon^2 \int_0^1 \tilde{\varphi}_n^{(k_,)} \sum_{l=1}^N F_{k_,}^{(l)} \, d\eta \tag{8.6.6}$$

$$\frac{dA_n^{(k_{,,})}}{dt} + \mu_n^{(k_{,,})} A_n^{(k_{,,})} = -\varepsilon^2 \int_0^1 \tilde{\varphi}_n^{(k_{,,})} \sum_{l=1}^N F_{k_{,,}}^{(l)} \, d\eta \tag{8.6.7}$$

For the homogeneous problem we have $F_{k_,}^{(l)} = 0(1)$ and $F_{k_{,,}}^{(l)} = 0(1)$. Moreover, $\tau_n^{(k)} = 0(1)$ positive for $n > 0$. As we have seen earlier under similar conditions, this permits us to conclude that

$$\left.\begin{aligned} A_n^{(k_,)} &= C_n^{(k_,)} e^{-\mu_n^{(k_,)} t} + 0(\varepsilon^2) \\ A_n^{(k_{,,})} &= C_n^{(k_{,,})} e^{-\mu_n^{(k_{,,})} t} + 0(\varepsilon^2) \end{aligned}\right\}; \quad n > 0 \tag{8.6.8}$$

It can easily be demonstrated that terms of order ε^2 in Eq. (8.6.8) are forced solutions which in the first approximation depend only on $A_0^{(k_,)}$ and $A_0^{(k_{,,})}$. The behaviour of perturbations is thus, in the final analysis, determined by the behaviour of the amplitude functions $A_0^{(k_,)}$ and $A_0^{(k_{,,})}$. Moreover, in calculating the forced solutions for $\psi_{\varepsilon\sigma}$ etc., the exponentially decaying terms of $A_n^{(k_,)}$, $A_n^{(k_{,,})}$ for $n > 0$ can again be disregarded on the same grounds as the series of exponential terms $\psi_k^{(l)}$ of Eq. (8.6.1).

We note finally that the conclusions concerning the role of exponential terms (8.6.1), (8.6.8) can also be deduced by a different argument, outlined in section 7.5. Thus we observe that the time scale of $A_0^{(k_,)}$ and $A_0^{(k_{,,})}$ is given by $\tau_0^{(k_,)} = 0(\varepsilon^2)$ and $\tau_0^{(k_{,,})} = 0(\varepsilon^2)$, while the time-scale of all terms in (8.6.1), (8.6.8) is of order of magnitude unity. Transforming the time variable in Eq. (8.6.6), (8.6.7) and performing the asymptotic expansion, the exponential terms disappear for all positive times.

8.7 Forced solutions for $\Psi_{\varepsilon\sigma}$ and $\Psi_{2k_0 \mp \varepsilon\sigma}$

We now turn to the effective calculation of the forced solutions and begin with the component $\psi_{\varepsilon\sigma}$. Utilizing Eqs. (8.6.5), (8.6.8) in Eq. (8.5.4) we obtain for $\psi_{\varepsilon\sigma}$ the non-homogeneous equation

$$\begin{aligned} \left[\mathfrak{L}_{\varepsilon\sigma} - \frac{\partial}{\partial t} S_{\varepsilon\sigma}\right] \psi_{\varepsilon\sigma} = \varepsilon \sum_{l=1}^N \{ & A_0^{(k_0)} \bar{A}_0^{(k_{,,})} [P_{k_0}^{(l)}(\varphi_0^{(k_0)}) \bar{Q}_{k_{,,}}(\bar{\varphi}_0^{(k_{,,})}) + \\ & + Q_{k_0}^{(l)}(\varphi_0^{(k_0)}) \bar{P}_{k_{,,}}^{(l)}(\bar{\varphi}_0^{(k_{,,})})] + \bar{A}_0^{(k_0)} A_0^{(k_,)} [\bar{P}_{k_0}^{(l)}(\bar{\varphi}_0^{(k_0)}) Q_{k_,}^{(l)}(\varphi_0^{(k_,)}) + \\ & + \bar{Q}_{k_0}^{(l)}(\bar{\varphi}_0^{(k_0)}) P_{k_,}^{(l)}(\varphi_0^{(k_,)})]\} + 0(\varepsilon^2) \end{aligned} \tag{8.7.1}$$

This is an equation of the type studied in section 6.5 and in fact the explicit solution (6.5.15) can be applied directly. However, the procedure leads to very cumbersome expressions if the complete form of Eq. (8.7.1) is retained. On the other hand, considerable simplification of Eq. (8.7.1) and its solution can be achieved within the context of the asymptotic theory, if certain auxiliary hypotheses are introduced.

Consider the functions $P_k^{(l)}(\varphi_0^{(k)})$ and $Q_k^{(l)}(\varphi_0^{(k)})$: we assume that these expressions are regular functions of the parameter k in the vicinity of $k = k_0$. If we now consider the values $k_{,} = k_0 + \varepsilon\,\sigma$ and $k_{,,} = k_0 - \varepsilon\,\sigma$. then a Taylor expansion leads directly to

$$\left.\begin{aligned} P_{k_{,}}^{(l)}(\varphi_0^{(k_{,})}) &= P_{k_0}^{(l)}(\varphi_0^{(k_0)}) + 0(\varepsilon) \\ Q_{k_{,}}^{(l)}(\varphi_0^{(k_{,})}) &= Q_{k_0}^{(l)}(\varphi_0^{(k_0)}) + 0(\varepsilon) \\ P_{k_{,,}}^{(l)}(\varphi_0^{(k_{,,})}) &= P_{k_0}^{(l)}(\varphi_0^{(k_0)}) + 0(\varepsilon) \\ Q_{k_{,,}}^{(l)}(\varphi_0^{(k_{,,})}) &= Q_{k_0}^{(l)}(\varphi_0^{(k_0)}) + 0(\varepsilon) \end{aligned}\right\} \tag{8.7.2}$$

with similar results for the complex conjugates. Substituting these developments into Eq. (8.7.1) we immediately find

$$\left[\mathfrak{L}_{\varepsilon\sigma} - \frac{\partial}{\partial t} S_{\varepsilon\sigma}\right]\psi_{\varepsilon\sigma} = \varepsilon\left[A_0^{(k_0)}\bar{A}_0^{(k_{,,})} + \bar{A}_0^{(k_0)}A_0^{(k_{,})}\right]\left[\sum_{l=1}^{N} f_0^{(l)}(\eta) + 0(\varepsilon)\right] \tag{8.7.3}$$

where $f_0^{(l)}$ is a function which has already been encountered in the theory of periodic solutions and is defined by Eq. (8.2.14).

Following section 6.5 we have

$$\psi_{\varepsilon\sigma}(\eta, t) = \sum_{n=0}^{\infty} A_n^{(\varepsilon\sigma)}(t)\,\varphi_n^{(\varepsilon\sigma)}(\eta) \tag{8.7.4}$$

$$A_n^{(\varepsilon\sigma)}(t) = -\varepsilon\, e^{-\mu_n^{(\varepsilon\sigma)}t}\left\{\int^{t} e^{\mu_n^{(\varepsilon\sigma)}t'}\left[A_0^{(k_0)}\bar{A}_0^{(k_{,,})} + \bar{A}_0^{(k_0)}A_0^{(k_{,})}\right]dt'\right\}\int_0^1 \tilde{\varphi}_n^{(\varepsilon\sigma)}\sum_{l=1}^{N} f_0^{(l)}\,d\eta \tag{8.7.5}$$

The result (8.7.5) can be further simplified in two ways; first we recall that

$$\frac{dA_0^{(k)}}{dt} = -i\,\nu_0^{(k)}A_0^{(k)} + 0(\varepsilon^2) \text{ for } \quad k = k_{,}; \quad k = k_{,,}; \quad k = k_0 \tag{8.7.6}$$

Integrating by parts in Eq. (8.7.5) we show that

$$\begin{aligned} &\int^{t} e^{\mu_n^{(\varepsilon\sigma)}(t'-t)}\left[A_0^{(k_0)}\bar{A}_0^{(k_{,,})} + \bar{A}_0^{(k_0)}A_0^{(k_{,})}\right]dt' \\ &\qquad = \frac{1}{\mu_n^{(\varepsilon\sigma)}}\left[A_0^{(k_0)}\bar{A}_0^{(k_{,,})} + \bar{A}_0^{(k_0)}A_0^{(k_{,})}\right] + 0(\varepsilon^2) \end{aligned} \tag{8.7.7}$$

We thus have

$$A_n^{(\varepsilon\sigma)} = -\left[A_0^{(k_0)}\bar{A}_0^{(k_{,,})} + \bar{A}_0^{(k_0)}A_0^{(k_{,})}\right]\frac{\varepsilon}{\mu_n^{(\varepsilon\sigma)}}\int_0^1 \sum_{l=1}^{N} f_0^{(l)}\,\tilde{\varphi}_n^{(\varepsilon\sigma)}\,d\eta \tag{8.7.8}$$

Final simplification is achieved by a generalization of the hypothesis of regularity with respect to k, so that by Taylor expansion

$$\frac{1}{\mu_n^{(\varepsilon\sigma)}}\int_0^1 \sum_{l=1}^N f_0^{(l)}\, \tilde{\varphi}_n^{(\varepsilon\sigma)}\, d\eta = \frac{1}{\mu_n^{(0)}}\int_0^1 \sum_{l=1}^N f_0^{(l)}\, \tilde{\varphi}_n^{(0)}\, d\eta + 0(\varepsilon) \qquad (8.7.9)$$

Introducing (8.7.8), (8.7.9) into solution (8.7.4), and comparing with Eqs. (8.2.7), (8.2.13) we conclude that the forced solution for $\psi_{\varepsilon\sigma}$ reads

$$\psi_{\varepsilon\sigma}(\eta, t) = \varepsilon\left[A_0^{(k_0)}\,\bar{A}_0^{(k_{\prime\prime})} + \bar{A}_0^{(k_0)}\,A_0^{(k_\prime)}\right]\left[g_0(\eta) + 0(\varepsilon)\right] \qquad (8.7.10)$$

where g_0 is a function already defined in the theory of periodic solutions and given by Eq. (8.2.13).

Turning to the forced solution for $\psi_{2k_0\mp\varepsilon\sigma}$ we can start from Eq. (8.5.5) and perform the calculation in exactly the same way as we just did for $\psi_{\varepsilon\sigma}$. The result is simply

$$\begin{aligned} \psi_{2k_0+\varepsilon\sigma} &= 2\varepsilon\, A_0^{(k_0)}\, A_0^{(k_\prime)}\left[g_2(\eta) + 0(\varepsilon)\right] \\ \psi_{2k_0-\varepsilon\sigma} &= 2\varepsilon\, A_0^{(k_0)}\, A_0^{(k_{\prime\prime})}\left[g_2(\eta) + 0(\varepsilon)\right] \end{aligned} \qquad (8.7.11)$$

where g_2 is again a function defined in Eq. (8.2.13).

The forced solutions for $\psi_{\varepsilon\sigma}$ and $\psi_{2k_0\pm\varepsilon\sigma}$ have thus all been expressed in terms of functions already defined in connection with periodic solutions, and it is this particular feature which permits us ultimately to arrive at simple results.

8.8 The equations for $A_0^{(k_\prime)}$ and $A_0^{(k_{\prime\prime})}$

We are now in a position to study the perturbations $\psi_{k_\prime}$ and $\psi_{k_{\prime\prime}}$ further. Returning to Eqs. (8.5.2), (8.5.3) we introduce the forced solutions (8.7.10), (8.7.11), while for the periodic solutions we utilize the results (8.2.7), (8.2.8), (8.2.9). We find

$$\begin{aligned} F_{k_\prime}^{(l)} = {} & |A_0^{(k_0)}|^2 A_0^{(k_\prime)}\{P_0^{(l)}(g_0)\, Q_{k_\prime}^{(l)}(\varphi_0^{(k_\prime)}) + Q_0^{(l)}(g_0)\, P_{k_\prime}^{(l)}(\varphi_0^{(k_\prime)}) + \\ & + 2\bar{P}_{k_0}^{(l)}(\bar{\varphi}_0^{(k_0)})\, Q_{2k_0+\varepsilon\sigma}^{(l)}(g_2) + 2\bar{Q}_{k_0}^{(l)}(\bar{\varphi}_0^{(k_0)})\, P_{2k_0+\varepsilon\sigma}^{(l)}(g_2)\} + \\ & + [A_0^{(k_0)}]^2\, \bar{A}_0^{(k_{\prime\prime})}\{P_{2k_0}^{(l)}(g_2)\, \bar{Q}_{k_{\prime\prime}}^{(l)}(\bar{\varphi}_0^{(k_{\prime\prime})}) + Q_{2k_0}^{(l)}(g_2)\, \bar{P}_{k_{\prime\prime}}^{(l)}(\bar{\varphi}_0^{(k_{\prime\prime})})\} + \\ & + \left[|A_0^{(k_0)}|^2 A_0^{(k_\prime)} + [A_0^{(k_0)}]^2\, \bar{A}_0^{(k_{\prime\prime})}\right]\{P_{k_0}^{(l)}(\varphi_0^{(k_0)})\, Q_{\varepsilon\sigma}^{(l)}(g_0) + \\ & + Q_{k_0}^{(l)}(\varphi_0^{(k_0)})\, P_{\varepsilon\sigma}^{(l)}(g_0)\} + 0(\varepsilon) \end{aligned} \qquad (8.8.1)$$

This result can considerably be simplified by the procedure utilized in the preceding section. In addition to Eq. (8.7.2) we introduce by Taylor expansion

$$\left.\begin{aligned} Q_{\varepsilon\sigma}^{(l)}(g_0) &= Q_0^{(l)}(g_0) + 0(\varepsilon) \\ Q_{2k_0+\varepsilon\sigma}^{(l)}(g_2) &= Q_{2k_0}^{(l)}(g_2) + 0(\varepsilon) \\ \text{etc.} & \end{aligned}\right\} \qquad (8.8.2)$$

Eq. (8.8.1) then becomes

$$F_{k_\prime}^{(l)} = \left\{2\,|A_0^{(k_0)}|^2 A_0^{(k_\prime)} + [A_0^{(k_0)}]^2 \bar{A}_0^{(k_{\prime\prime})}\right\}\left\{f_1^{(l)}(\eta) + 0(\varepsilon)\right\} \tag{8.8.3}$$

where $f_1^{(l)}$ is the function defined in Eq. (8.2.12). In a similar way we also obtain

$$F_{k_{\prime\prime}}^{(l)} = \left\{2\,|A_0^{(k_0)}|^2 A_0^{(k_{\prime\prime})} + [A_0^{(k_0)}]^2 \bar{A}_0^{(k_\prime)}\right\}\left\{f_1^{(l)}(\eta) + 0(\varepsilon)\right\} \tag{8.8.4}$$

From the eigenfunction expansion of the perturbations $\psi_{k_\prime}$ and $\psi_{k_{\prime\prime}}$ by Eqs. (8.6.5), (8.6.6), (8.6.7) we now find the following two equations for the amplitude functions $A_0^{(k_\prime)}$ and $A_0^{(k_{\prime\prime})}$:

$$\frac{dA_0^{(k_\prime)}}{dt} + \mu_0^{(k_\prime)} A_0^{(k_\prime)} = -\varepsilon^2\left[2\,|A_0^{(k_0)}|^2 A_0^{(k_\prime)} + [A_0^{(k_0)}]^2 \bar{A}_0^{(k_{\prime\prime})}\right]\left[\int_0^1 \tilde{\varphi}_0^{(k_\prime)} \sum_{l=1}^{N} f_1^{(l)}\, d\eta + 0(\varepsilon)\right] \tag{8.8.5}$$

$$\frac{dA_0^{(k_{\prime\prime})}}{dt} + \mu_0^{(k_{\prime\prime})} A_0^{(k_{\prime\prime})} = -\varepsilon^2\left[2\,|A_0^{(k_0)}|^2 A_0^{(k_{\prime\prime})} + [A_0^{(k_0)}]^2 \bar{A}_0^{(k_\prime)}\right]\left[\int_0^1 \tilde{\varphi}_0^{(k_{\prime\prime})} \sum_{l=1}^{N} f_1^{(l)}\, d\eta + 0(\varepsilon)\right] \tag{8.8.6}$$

Final simplification is achieved by introduction of the developments

$$\int_0^1 \tilde{\varphi}_0^{(k_\prime)} \sum_{l=1}^{N} f_1^{(l)}\, d\eta = \int_0^1 \tilde{\varphi}_0^{(k_0)} \sum_{l=1}^{N} f_1^{(l)}\, d\eta + 0(\varepsilon) = \beta_0 + 0(\varepsilon) \tag{8.8.7}$$

$$\int_0^1 \tilde{\varphi}_0^{(k_{\prime\prime})} \sum_{l=1}^{N} f_1^{(l)}\, d\eta = \int_0^1 \tilde{\varphi}_0^{(k_0)} \sum_{l=1}^{N} f_1^{(l)}\, d\eta + 0(\varepsilon) = \beta_0 + 0(\varepsilon) \tag{8.8.8}$$

On the right-hand-side of Eqs. (8.8.7), (8.8.8) we again find the constant β_0 which played an essential role in the theory of periodic solutions. Finally we have for the amplitude functions $A_0^{(k_\prime)}$ and $A_0^{(k_{\prime\prime})}$ the two coupled equations

$$\frac{dA_0^{(k_\prime)}}{dt} + \mu_0^{(k_\prime)} A_0^{(k_\prime)} = -\varepsilon^2\left[2\,|A_0^{(k_0)}|^2 A_0^{(k_\prime)} + [A_0^{(k_0)}]^2 \bar{A}_0^{(k_{\prime\prime})}\right][\beta_0 + 0(\varepsilon)] \tag{8.8.9}$$

$$\frac{dA_0^{(k_{\prime\prime})}}{dt} + \mu_0^{(k_{\prime\prime})} A_0^{(k_{\prime\prime})} = -\varepsilon^2\left[2\,|A_0^{(k_0)}|^2 A_0^{(k_{\prime\prime})} + [A_0^{(k_0)}]^2 \bar{A}_0^{(k_\prime)}\right][\beta_0 + 0(\varepsilon)] \tag{8.8.10}$$

8.9 Solution of the stability problem for $k_0 = k_{cr}$

As we have seen in section 8.2, in the presence of perturbations of the level considered here, the components of the periodic solution tend to solutions harmonic in time in the limit $t \to \infty$. Consequently, we write

$$A_0^{(k_0)}(t) = a_0^{(k_0)} e^{i\omega t}[1 + 0(\varepsilon)] \tag{8.9.1}$$

where

$$a_0^{(k_0)} = \mp [Re(\beta_0)]^{-1/2} + 0(\varepsilon^2) \tag{8.9.2}$$

$$\omega = -\nu_0^{(k_0)} + 0(\varepsilon^2) \tag{8.9.3}$$

Inspection of Eqs. (8.8.9), (8.8.10) shows that under those circumstances the solutions for $A_0^{(k_\prime)}$ and $A_0^{(k_{\prime\prime})}$ must be of the form

$$\begin{aligned} A_0^{(k_\prime)}(t) &= e^{i\omega_\prime t} a_0^{(k_\prime)}(t) \\ A_0^{(k_{\prime\prime})}(t) &= e^{i\omega_{\prime\prime} t} a_0^{(k_{\prime\prime})}(t) \end{aligned} \tag{8.9.4}$$

where $a_0^{(k_\prime)}$ and $a_0^{(k_{\prime\prime})}$ are real functions of the time variable. Moreover, inserting (8.9.4) into Eqs. (8.8.9), (8.8.10) and separating the real and the imaginary parts we find that

$$\begin{aligned} \omega_\prime &= -\nu_0^{(k_0)} - \varepsilon\sigma\left(\frac{\partial \nu_0^{(k)}}{\partial k}\right)_{k=k_0} + 0(\varepsilon^2) \\ \omega_{\prime\prime} &= -\nu_0^{(k_0)} + \varepsilon\sigma\left(\frac{\partial \nu_0^{(k)}}{\partial k}\right)_{k=k_0} + 0(\varepsilon^2) \end{aligned} \tag{8.9.5}$$

while $a_0^{(k_\prime)}$ and $a_0^{(k_{\prime\prime})}$ must satisfy the equations

$$\begin{aligned} \frac{d a_0^{(k_\prime)}}{dt} + \tau_0^{(k_\prime)} a_0^{(k_\prime)} &= -\varepsilon^2\{2|a_0^{(k_0)}|^2 a_0^{(k_\prime)} + |a_0^{(k_0)}|^2 a_0^{(k_{\prime\prime})}\}[Re(\beta_0) + 0(\varepsilon)] \\ \frac{d a_0^{(k_{\prime\prime})}}{dt} + \tau_0^{(k_{\prime\prime})} a_0^{(k_{\prime\prime})} &= -\varepsilon^2\{2|a_0^{(k_0)}|^2 a_0^{(k_{\prime\prime})} + |a_0^{(k_0)}|^2 a_0^{(k_\prime)}\}[Re(\beta_0) + 0(\varepsilon)] \end{aligned} \tag{8.9.6}$$

Using Eq. (8.9.2) and recalling that in the present case $\varepsilon^2 = -\tau_0^{(k_0)}$ we finally arrive at the system of equations

$$\begin{aligned} \frac{d a_0^{(k_\prime)}}{dt} + \tau_0^{(k_\prime)} a_0^{(k_\prime)} &= \tau_0^{(k_0)}[2a_0^{(k_\prime)} + a_0^{(k_{\prime\prime})}][1 + 0(\varepsilon)] \\ \frac{d a_0^{(k_{\prime\prime})}}{dt} + \tau_0^{(k_{\prime\prime})} a_0^{(k_{\prime\prime})} &= \tau_0^{(k_0)}[2a_0^{(k_{\prime\prime})} + a_0^{(k_\prime)}][1 + 0(\varepsilon)] \end{aligned} \tag{8.9.7}$$

The solutions are obviously exponential functions, which can be determined by writing

$$\begin{aligned} a_0^{(k_\prime)} &= C_1' e^{-\lambda_1 t} + C_2' e^{-\lambda_2 t} \\ a_0^{(k_{\prime\prime})} &= C_1'' e^{-\lambda_1 t} + C_2'' e^{-\lambda_2 t} \end{aligned} \tag{8.9.8}$$

For the constants C_1', C_2', C_1'', C_2'' we have, first, from initial conditions

$$\begin{aligned} C_1' + C_2' &= C_0^{(k_\prime)} = \text{a given initial value} \\ C_1'' + C_2'' &= C_0^{(k_{\prime\prime})} = \text{a given initial value} \end{aligned} \tag{8.9.9}$$

Then introducing Eq. (8.9.8) into (8.9.7) we have the relations

$$\tau_0^{(k_0)} C_1' = [-\lambda_1 + \tau_0^{(k_{\prime\prime})} - 2\tau_0^{(k_0)}] C_1'' \tag{8.9.10}$$

$$\tau_0^{(k_0)} C_2' = [-\lambda_2 + \tau_0^{(k_\prime)} - 2\tau_0^{(k_0)}] C_2'' \tag{8.9.11}$$

Finally, λ_1 and λ_2 must be solutions of the equation

$$\lambda^2 - [\tau_0^{(k_\prime)} + \tau_0^{(k_{\prime\prime})} - 4\tau_0^{(k_0)}]\,\lambda + [(\tau_0^{(k_\prime)} - 2\tau_0^{(k_0)})\,(\tau_0^{(k_{\prime\prime})} - 2\tau_0^{(k_0)}) - (\tau_0^{(k_0)})^2] = 0 \tag{8.9.12}$$

that is

$$\lambda_{1,2} = -2\tau_0^{(k_0)} + \tfrac{1}{2}(\tau_0^{(k_\prime)} + \tau_0^{(k_{\prime\prime})}) \mp \sqrt{[\tau_0^{(k_0)}]^2 + \tfrac{1}{4}[\tau_0^{(k_\prime)} - \tau_0^{(k_{\prime\prime})}]^2} \tag{8.9.13}$$

It is at this stage that a representation of $\tau_0^{(k)}$ as a function of k is needed. In section 8.3 we have found that in the region $k - k_{cr} = 0(\varepsilon)$ we may use the approximation

$$\tau_0^{(k)} = \tau_0^{(k_0)}\{1 - \alpha_1\gamma_2\sigma^2 + 0(\varepsilon)\} \tag{8.9.14}$$

where $k_0 = k_{cr}$, while α_1 and γ_2 are positive constants. Substituting this approximation for $\tau_0^{(k_\prime)}$ and $\tau_0^{(k_{\prime\prime})}$ we obtain

$$\begin{aligned} \lambda_1 &= -\tau_0^{(k_0)}\alpha_1\gamma_2\sigma^2 \\ \lambda_2 &= -\tau_0^{(k_0)}[2 + \alpha_1\gamma_2\sigma^2] \end{aligned} \tag{8.9.15}$$

Since now $\tau_0^{(k_0)} < 0$ we have $\lambda_1 >$ and $\lambda_2 > 0$.

Consequently, from formula (8.8.9) it follows that

$$\lim_{t\to\infty} A_0^{(k_\prime)} = 0 \quad \text{and} \quad \lim_{t\to\infty} A_0^{(k_{\prime\prime})} = 0 \tag{8.9.16}$$

In the result (8.9.16) we have the solution of the stability problem. All other quantities describing the perturbations can now be calculated explicitly; their behaviour being determined by the behaviour of $A_0^{(k_\prime)}$ and $A_0^{(k_{\prime\prime})}$ one finds without difficulty that

$$\lim_{t\to\infty} A_n^{(k_\prime)} = 0; \qquad \lim_{t\to\infty} A_n^{(k_{\prime\prime})} = 0 \tag{8.9.17}$$

and finally

$$\begin{aligned} &\lim_{t\to\infty} \psi_{k_\prime} = 0; \qquad \lim_{t\to\infty} \psi_{k_{\prime\prime}} = 0 \\ &\lim_{t\to\infty} \psi_{\varepsilon\sigma} = 0; \qquad \lim_{t\to\infty} \psi_{2k_0 \mp \varepsilon\sigma} = 0 \end{aligned} \tag{8.9.18}$$

etc.

Thus the periodic solution of period $2\pi/k_{cr}$ is found to be stable with respect to small aperiodic perturbations Φ_k, $k \neq m\,k_{cr}$.

8.10 Regions of validity of the asymptotic results

It should be noted that in the analysis of the preceding sections certain perturbations have not yet been considered. These perturbations do not interact directly with the perturbations which have been studied, but their behaviour must still be investigated. In section 8.4 we have considered perturbations Φ_k corresponding to k values such that $\tau_0^{(k)} = 0(1)$, while in sections 8.5 and following the values of k ware chosen so that $\tau_0^{(k)} = 0(\varepsilon^2)$. Looking over fig. 11 it becomes clear that the whole range of k values has not yet been covered and in fact we need a continuation of the asymptotic results to obtain validity in the domain of k values corresponding to say $\tau_0^{(k)} = 0(\varepsilon^{\alpha})$, where $0 < \alpha < 2$. We now outline how this extension may be obtained.

All the results of sections 8.5, 8.6 hold formally in the domain of values of k that we wish to study now, provided that for the quantity $\varepsilon\sigma$ we choose values such as to make $k_{,} = k_0 + \varepsilon\sigma$ and $k_{,,} = k_0 - \varepsilon\sigma$ lie in this domain. Taking up the results of these sections we consider Eqs. (8.6.6), (8.6.7). If $\tau_0^{(k)} = 0(\varepsilon^{\alpha})$, with $0 < \alpha < 2$, the asymptotic solution of these equations for $n = 0$ can be shown to be given by

$$\begin{aligned} A_0^{(k_{,})} &= C_0^{(k_{,})} e^{-\mu_0^{(k_{,})} t} + 0(\varepsilon^{2-\alpha}) \\ A_0^{(k_{,,})} &= C_0^{(k_{,,})} e^{-\mu_0^{(k_{,,})} t} + 0(\varepsilon^{2-\alpha}) \end{aligned} \tag{8.10.1}$$

where terms of order $\varepsilon^{2-\alpha}$ are forced solutions. Since in the domain of present interest $\tau_0^{(k)}$ is positive, we see that $A_0^{(k_{,})}$, $A_0^{(k_{,,})}$ tend to zero for $t \to \infty$. Starting from Eq. (8.10.1) all other quantities can be evaluated and the conclusion is reached that Φ_k, and all perturbation-components that interact with it, decay to zero as $t \to \infty$.

On the other hand, the solution obtained in section 8.9 can be shown to hold in an extended domain, defined by $\varepsilon\sigma = 0(\varepsilon^{\gamma})$ with $0 < \gamma \leqq 1$. This extended domain can be shown to overlap with the domain of validity of the solution (8.10.1). Finally, the two solutions can be shown to be identical in the common region of validity.

We consider now the validity of the results of the preceding sections in the case when k takes values extremely close to k_{cr}. A careful analysis of the order of magnitude of terms that have been neglected shows that in the solution for λ_1, Eq. (8.9.15), we have

$$\lambda_1 = -\tau_0^{(k_{cr})} \{(\alpha_1 \gamma_2)\, \sigma^2 + 0(\varepsilon\sigma) + 0(\varepsilon^2)\} \tag{8.10.2}$$

When σ takes very small values, all terms in Eq. (8.10.2) become of the same order of magnitude. Thus, for $\sigma = 0(\varepsilon)$, Eq. (8.10.2) only shows that $\lambda_1 = 0(\varepsilon^4)$, but no conclusion concerning the sign of λ_1 can be drawn. It follows that in the region $k - k_{cr} = 0(\varepsilon)$ we must distinguish a subregion $k - k_{cr} = 0(\varepsilon^2)$; the asymptotic theory that we have developed does not permit as to determine the behaviour of the perturbations Φ_k if $k - k_{cr} = 0(\varepsilon^2)$. In fact, it can be shown that when one considers this sub-region, non-linear terms in the equations governing the perturbation Φ_k must be taken into account.

8.11 Stability of periodic solutions in the case $k_0 \neq k_{cr}$

Up to now we have studied the stability of a particular periodic solution, corresponding to $k_0 = k_{cr}$. Presently we generalize our results and consider periodic solutions corresponding to other values of k_0.

Re-considering the analysis of this chapter one easily sees that most of it remains valid for arbitrary values of k_0. In fact, it is only after the solution (8.9.13) has been obtained, that the choise of the k_0 value must be specified. We thus take up Eq. (8.9.13), which reads

$$\lambda_{1.2} = -2\tau_0^{(k_0)} + \tfrac{1}{2}\left(\tau_0^{(k_,)} + \tau_0^{(k_{,,})}\right) \pm \sqrt{[\tau_0^{(k_0)}]^2 + \tfrac{1}{4}[\tau_0^{(k_,)} - \tau_0^{(k_{,,})}]^2} \quad (8.11.1)$$

$$k_, = k_0 + \varepsilon\,\sigma; \quad k_{,,} = k_0 - \varepsilon\,\sigma; \quad \varepsilon^2 = |\tau_0^{(k_0)}| \quad (8.11.2)$$

On the other hand, the approximation for $\tau_0^{(k)}$, developed in section 8.3, is entirely independent of k_0. In order to consider the case $k_0 \neq k_{cr}$, a slightly modified notation is necessary. We write

$$k = k_{cr} + \sqrt{|\tau_0^{(k_{cr})}|}\,\bar{\sigma} \quad (8.11.3)$$

The parabolic approximation for $\tau_0^{(k)}$ then reads

$$\tau_0^{(k)} = \tau_0^{(k_{cr})}\{1 - (\alpha_1\,\gamma_2)\,\bar{\sigma}^2\} \quad (8.11.4)$$

k_0 is defined by

$$k_0 = k_{cr} + \sqrt{|\tau_0^{(k_{cr})}|}\,\bar{\sigma}_0$$

and similarly

$$\left.\begin{aligned} k_, &= k_{cr} + \sqrt{|\tau_0^{(k_{cr})}|}\,\bar{\sigma}_, \\ k_{,,} &= k_{cr} + \sqrt{|\tau_0^{(k_{cr})}|}\,\bar{\sigma}_{,,} \end{aligned}\right\} \quad (8.11.5)$$

Comparing with Eq. (8.11.2) we have

$$\bar{\sigma}_, = \bar{\sigma}_0 + \sqrt{\left|\frac{\tau_0^{(k_0)}}{\tau_0^{(k_{cr})}}\right|}\,\sigma; \quad \bar{\sigma}_{,,} = \bar{\sigma}_0 - \sqrt{\left|\frac{\tau_0^{(k_0)}}{\tau_0^{(k_{cr})}}\right|}\,\sigma \quad (8.11.6)$$

so that, using formula (8.11.4) for $\bar{\sigma} = \bar{\sigma}_0$, $\bar{\sigma} = \bar{\sigma}_,$ and $\bar{\sigma} = \bar{\sigma}_{,,}$ we obtain the approximations

$$\left.\begin{aligned} \tau_0^{(k_0)} &= \tau_0^{(k_{cr})}\{1 - \alpha_1\,\gamma_2\,\bar{\sigma}_0^2\} \\ \tau_0^{(k_,)} &= \tau_0^{(k_{cr})}\{1 - \alpha_1\gamma_2[\bar{\sigma}_0 + \sqrt{1 - \alpha_1\,\gamma_2\,\bar{\sigma}_0^2}\,\sigma]^2\} \\ \tau_0^{(k_{,,})} &= \tau_0^{(k_{cr})}\{1 - \alpha_1\gamma_2[\bar{\sigma}_0 - \sqrt{1 - \alpha_1\,\gamma_2\,\bar{\sigma}_0^2}\,\sigma]^2\} \end{aligned}\right\} \quad (8.11.7)$$

Introducing result (8.11.7) into the solution (8.11.1) we finally have

$$\lambda_1 = |\tau_0^{(k_{cr})}|\,(1 - \alpha_1\,\gamma_2\,\bar{\sigma}_0^2)\left\{1 + \alpha_1\gamma_2\,\sigma^2 - \sqrt{1 + \frac{4\,(\alpha_1\,\gamma_2)^2\,\bar{\sigma}_0^2\,\sigma^2}{1 - \alpha_1\,\gamma_2\,\bar{\sigma}_0^2}}\right\} \quad (8.11.8)$$

A periodic solution corresponding to a fixed value of k_0 (and hence a fixed value of $\bar{\sigma}_0$) is stable with respect to perturbations Φ_k, $k \neq m\,k_0$, if for that value of $\bar{\sigma}_0$, λ_1 is positive for all values of σ.

Analysing Eq. (8.11.8) we find in the first place that

$$\left.\begin{aligned} &[\lambda_1]_{\sigma=0} = 0; \quad \left(\frac{\partial\,\lambda_1}{\partial\sigma}\right)_{\sigma=0} = 0 \\ &\left(\frac{\partial^2\,\lambda_1}{\partial\sigma^2}\right)_{\sigma=0} = |\tau_0^{(k_{cr})}|\,2\alpha_1\,\gamma_2[1 - 3\alpha_1\,\gamma_2\,\bar{\sigma}_0^2] \end{aligned}\right. \quad (8.11.9)$$

secondly, we also have

$$\lambda_1 = 0 \quad \text{if} \quad \sigma^2 = \frac{6}{1 - \alpha_1 \gamma_2 \bar{\sigma}_0^2} \left[\bar{\sigma}_0^2 - \frac{1}{3\alpha_1 \gamma_2} \right] \tag{8.11.10}$$

Plotting now λ_1 as a function of σ, or, more conveniently, as a function of $\bar{\sigma}_1$, we obtain the result indicated in fig. 12.

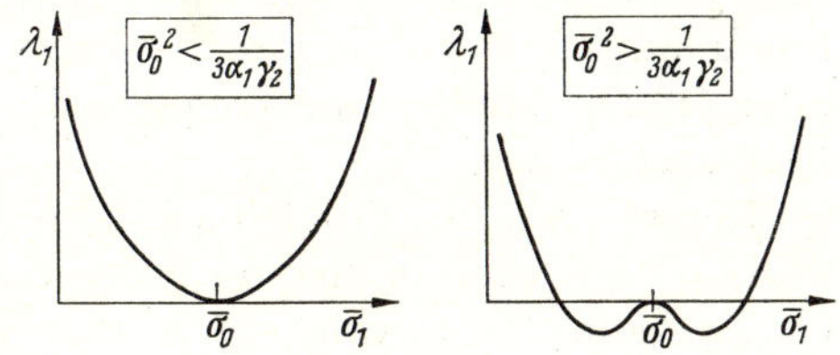

Figure 12

We see that λ_1 is positive if $\bar{\sigma}_0^2 < (3\alpha_1 \gamma_2)^{-1}$. Consequently, the solution of period $2\pi/k_0$ is stable if k_0 satisfies the condition

$$|k_0 - k_{cr}| < \sqrt{\frac{|\tau_0^{(k_{cr})}|}{3\alpha_1 \gamma_2}} \tag{8.11.11}$$

Using Eq. (8.3.4), this condition can finally be written in the form

$$|k_0 - k_{cr}| < \sqrt{\frac{|R - R_{cr}|}{3\gamma_2}} \tag{8.11.12}$$

8.12 Summary and interpretation of results

Let us recall the principal features of the problem studied in this chapter. We have considered the case of weak instability, with the parameter R having a fixed value $R > R_{cr}$, chosen so that $|R - R_{cr}|$ is small (more precisely, $|R - R_{cr}| = 0(\varepsilon^2)$). There exists, in that case, an interval of values of the parameter k defined by

$$|k - k_{cr}| < \sqrt{\frac{|R - R_{cr}|}{\gamma_2}} \tag{8.12.1}$$

which may be called the interval of instability. In the linearized theory, a perturbation Φ_k grows without limit as a function of time when k is in that interval. In non-linear theory (chapter 7), when certain conditions are satisfied, there exists for every $k = k_0$ verifying relation (8.12.1) a periodic solution with period $2\pi/k_0$ which remains bounded and of order ε, and which tends to a solution harmonic in time in the limit as $t \to \infty$.

Studying the stability of such a periodic solution with respect to small aperiodic perturbations Φ_k, $k \neq m\, k_0$, we have found that in the interval of *possible* periodic solutions (8.12.1) there exists a subinterval of *stable* periodic solutions, defined by

$$|k_0 - k_{cr}| < \sqrt{\frac{|R - R_{cr}|}{3\gamma_2}} \tag{8.12.2}$$

If k_0 is in the interval (8.12.2), then

$$\lim_{t \to \infty} \Phi_k = 0 \quad \text{for} \quad k \neq m\, k_0, \quad m = 0, 1, 2, \ldots \tag{8.12.3}$$

Upon some reflection, this result appears quite remarkable. For every choice of k_0 satisfying Eq. (8.12.2), the perturbations Φ_k, $k \neq m k_0$, include components contained in the unstable interval (8.12.1). These 'unstable perturbations', which grow without bounds in the linearized theory, are completely stabilized in the presence of the periodic solution and tend to zero as $t \to \infty$. Of all the effects of non-linear interactions, studied in the preceding chapters, this is perhaps the most spectacular.

Returning to the problem of stability of a basic solution φ_0, formulated and discussed in chapter 6, we can state that a periodic solution corresponding to a value of k that satisfies the condition (8.12.2) can be interpreted as a limit for $t \to \infty$ of an initially aperiodic perturbation $\Phi'(\xi, \eta, 0)$. However, every stable periodic solution represents a limit, so that k_0 can yet take any value satisfying relation (8.12.2). Moreover, the stable periodic solutions can only be considered as *possible* limits, at this stage. In other words, given an arbitrary initial perturbation $\Phi'(\xi, \eta, 0)$, we cannot yet say under what conditions the perturbation will tend in the limit $t \to \infty$ to one of the stable periodic solutions, nor can we decide which periodic solution will be reached in the limit. The problem that arises now can be formulated as follows.

An arbitrary aperiodic perturbation $\Phi'(\xi, \eta, 0)$ provides the initial values of the components Φ_k by formula (6.4.2). We wish to determine, for every value of k_0 satisfying relation (8.12.2), the class of all initial values $\Phi_k(\eta, 0)$, for which the periodic solution corresponding to k_0 represents the limit of the perturbations as $t \to \infty$.

From the analysis of this chapter we can conclude that for every k_0 satisfying (8.12.2) such a class of initial values exists. In fact, the perturbations defined in Eqs. (8.2.1), (8.2.2), (8.2.3) can be interpreted as examples of initial values belonging to the classes that we would like to determine. But to arrive at the desired classification we need to take up the problem of stability of periodic solutions and study it in a non-linear approximation.

Thus it is necessary to study the stability with respect to perturbations Φ_k, $m \neq m k_0$, whose level is higher than that defined in Eq. (8.2.2). On the basis of analogous but simpler non-linear problems studied in the preceding chapters, we can expect that critical values for the perturbations Φ_k will exist, limiting the domain of initial values for which Φ_k, $k \neq m k_0$, tend to zero as $t \to \infty$.

It is in this perspective that the correlation of classes of initial values and stable periodic solutions of different periods can be achieved, but the non-linear problem that has to be solved for this purpose appears as one of extreme complexity.

Chapter 9

PERIODIC SOLUTIONS IN POISEUILLE FLOW

9.1 Introduction

As we already have remarked, the analysis of stability of Poiseuille flow leads to a mathematical problem which belongs to the class defined and studied in chapters 6 and 7. A separate and somewhat more detailed investigation of this problem appears of interest for two reasons.

Unlike the mathematical models studied in chapter 5 and in section 7.7, the analysis of Poiseuille flow provides an example of application of the theory to a stability problem which arises in the physical world. It is true that two-dimensional Poiseuille flow cannot exactly be realized experimentally and two-dimensional perturbations cannot be imposed. Nevertheless, in its mathematical aspects, the two-dimensional Poiseuille flow may be considered as the first step in the study of more realistic problems of hydrodynamic stability.

Furthermore, the non-linear behaviour of periodic perturbations in Poiseuille flow has been the subject of various investigations, culminating in the analysis of STUART (1960b) and WATSON (1960). By applying the present method we have thus an opportunity to compare its results to those of earlier investigations.

The history of research on non-linear stability of two-dimensional Poiseuille flow can be found described in STUART's (1960b) paper, but the significant advance in the problem is due to Stuart himself. Inspired by LANDAU's (1944) conjecture, Stuart first deduced a 'Landau-type' equation (see section 3.2) in his 1958 paper, by a method based on energy balance considerations. But as the approximations involved were not too satisfactory, in STUART (1960b) the differential equations of the problem are attacked directly. Periodic solutions are studied, and by approximations based on order of magnitude considerations the infinite system of equations is reduced to a finite one. Inspection shows the result to be roughly equivalent to Eqs. (7.1.5), (7.2.6), (7.2.7), (7.3.4) of the present asymptotic method. In solving the resulting system of equations Stuart leans on WATSON's (1960) analysis, which is meant to provide the mathematical foundation of Stuart's theory. Anticipating somewhat, we can state that application of the theory of chapters 6 and 7 to the Poiseuille flow problem, which will be elaborated below, leads to results formally identical with those of STUART (1960b).

WATSON's (1960) study is concerned with series expansions, so constructed that Stuart's theory appears as an approximation in which only the first terms of the series have been retained. The concept of eigenfunction expansion is not used, although at closer analysis several essential steps in Stuart's and Watson's method appear difficult to justify without that concept. Also, among the series expansions introduced by Watson, several are meaningfull only if considered as what we have called 'forced solutions'. The adjoint linearized problem has been introduced into the theory by Stuart, and is utilized in Watson's analysis, where it serves as a means of writing solutions in a compact closed form. However, the orthogonality relations are neither defined nor discussed, and their fundamental role does not seem to have been fully realized.

These hidden aspects of the theory ware understood to some extend by SEGEL (1962), who attempted to explain the Stuart-Watson method, in connection with a thermal instability problem, by analogy with the eigenfunction expansion method for solution of non-homogeneous differential equations. However, Segel did not pursue the matter beyond the stage of general argument.

A different approach has recently been proposed by STRUMINSKI (1963b) who used Poincaré's method, but does not appear to have succeded in defining correctly the small parameters of the problem.

Stuart's and Watson's theory is concerned with formal analysis of periodic solutions, as is chapter 7 in the present study. The question of stability of periodic solutions with respect to *aperiodic* perturbations, essential for the interpretation of the periodic solutions, has not been investigated so far. Our chapter 8 is applicable here and we shall have the opportunity to discuss its results. Let us mention in this connection that other types of interaction of different perturbations have been investigated by GÖRTLER and WITTING (1957), BENNEY and LIN (1960)[1] and by STUART (1961). In those studies a two-dimensional unstable perturbation is considered to interact with a three-dimensional periodic perturbation in a two-dimensional flow. This results in a multitude of non-linear effects, some of which are emphasized in BENNEY's (1961) calculations, while others are discussed more in detail by STUART (1961).

In what follows, we develop the theory of periodic solutions in Poiseuille flow along the lines of chapters 6 and 7. The interpretation of the results on the basis of the analysis of chapter 8 is considered in the final section of the chapter.

---

[1] Those studies are strictly speaking not concerned with Poiseuille flow, but with a parallel-flow approximation of the boundary layer. The problem becomes then mathematicaly very similar to that of Poiseuille flow, but in the context of the non-linear theory the approximation is only acceptable as a mathematical model.

9.2 Formulation of the stability problem

We recall that the stream function $\Phi(\xi, \eta, t)$ in a two-dimensional Poiseuille flow satisfies the equation

$$\frac{1}{R}\nabla^2(\nabla^2\Phi) - \frac{\partial}{\partial t}\nabla^2\Phi = \frac{\partial\Phi}{\partial\eta}\frac{\partial\nabla^2\Phi}{\partial\xi} - \frac{\partial\Phi}{\partial\xi}\frac{\partial\nabla^2\Phi}{\partial\eta} \qquad (9.2.1)$$

with boundary conditions

$$\left(\frac{\partial\Phi}{\partial\eta}\right)_{\eta=0} = \left(\frac{\partial\Phi}{\partial\eta}\right)_{\eta=1} = \left(\frac{\partial\Phi}{\partial\xi}\right)_{\eta=0} = \left(\frac{\partial\Phi}{\partial\xi}\right)_{\eta=1} = 0 \qquad (9.2.2)$$

where R is the Reynolds number, and $\nabla^2 = \frac{\partial^2}{\partial\xi^2} + \frac{\partial^2}{\partial\eta^2}$.

We wish to study the stability of the basic solution

$$\Phi = \varphi_0(\eta) = \tfrac{1}{2}\eta^2 - \tfrac{1}{3}\eta^3 + \text{cont.} \qquad (9.2.3)$$

For this purpose we write

$$\Phi(\xi, \eta, t) = \varphi_0(\eta) + \Phi'(\xi, \eta, t) \qquad (9.2.4)$$

where Φ' is the perturbation. Introducing into Eq. (9.2.1) we find that the perturbation must satisfy the equation

$$\frac{1}{R}\nabla^2(\nabla^2\Phi') - \frac{\partial\nabla^2\Phi'}{\partial t} - \frac{d\varphi_0}{d\eta}\frac{\partial\nabla^2\Phi'}{\partial\xi} + \frac{d^3\varphi_0}{d\eta^3}\frac{\partial\Phi'}{\partial\xi}$$
$$= \frac{\partial\Phi'}{\partial\eta}\frac{\partial\nabla^2\Phi'}{\partial\xi} - \frac{\partial\Phi'}{\partial\xi}\frac{\partial\nabla^2\Phi'}{\partial\eta} \qquad (9.2.5)$$

with boundary conditions (9.2.2).

In contrast to chapter 6, we pass directly to the periodic solutions, since their interpretation is now available from the general theory of chapter 6 and 8. We write

$$\Phi'(\xi, \eta, t) = \sum_{m=-\infty}^{m=+\infty} e^{-ik_0 m\xi}\,\Phi_m(\eta, t) \qquad (9.2.6)$$

and obtain the system of equations

$$\mathfrak{L}_m(\Phi_m) - \frac{\partial}{\partial t}S_m(\Phi_m) = F_m^{(1)} + F_m^{(2)} \qquad (9.2.7)$$

The operators $\mathfrak{L}_m$ and S_m are defined by

$$S_m = \frac{\partial^2}{\partial\eta^2} - m^2 k_0^2 \qquad (9.2.8)$$

$$\mathfrak{L}_m = \frac{1}{R}S_m^2 + i\,m\,k_0\left[\frac{d\varphi_0}{d\eta}S_m - \frac{d^3\varphi_0}{d\eta^3}\right] \qquad (9.2.9)$$

The non-linear terms are given by

$$F_m^{(1)} + F_m^{(2)} = -i\,m\,k_0 \left\{ \frac{\partial \Phi_0}{\partial \eta} S_m(\Phi_m) - \frac{\partial^3 \Phi_0}{\partial \eta^3} \Phi_m \right\} +$$

$$+ i\,k_0 \sum_{n=1}^{m-1} \left\{ n\,\Phi_n \frac{\partial}{\partial \eta} S_{m-n}(\Phi_{m-n}) - (m-n) \frac{\partial \Phi_n}{\partial \eta} S_{m-n}(\Phi_{m-n}) \right\} -$$

$$- i\,k_0 \sum_{n=1}^{\infty} \left\{ n\,\bar{\Phi}_n \frac{\partial}{\partial \eta} S_{m+n}(\Phi_{m+n}) - (m+n)\,\Phi_{m+n} \frac{\partial}{\partial \eta} S_n(\bar{\Phi}_n) + \right.$$

$$\left. + (m+n) \frac{\partial \bar{\Phi}_n}{\partial \eta} S_{m+n}(\Phi_{m+n}) - n \frac{\partial \Phi_{m+n}}{\partial \eta} S_n(\bar{\Phi}_n) \right\} \tag{9.2.10}$$

Moreover, every component Φ_m, $m \neq 0$, must satisfy the boundary conditions

$$(\Phi_m)_{\eta=0} = (\Phi_m)_{\eta=1} = \left(\frac{\partial \Phi_m}{\partial \eta}\right)_{\eta=0} = \left(\frac{\partial \Phi_m}{\partial \eta}\right)_{\eta=1} = 0 \tag{9.2.11}$$

Concerning the component Φ_0 a few additional remarks are in order. We recall that Φ_0 represents the modification of the basic flow due to presence of the perturbations. Application of boundary conditions (9.2.2) to the series (9.2.6) leads to only two conditions for Φ_0:

$$\left(\frac{\partial \Phi_0}{\partial \eta}\right)_{\eta=0} = \left(\frac{\partial \Phi_0}{\partial \eta}\right)_{\eta=1} = 0 \tag{9.2.12}$$

On the other hand, Eqs. (9.2.7) to (9.2.10), for $m = 0$, simplify to

$$\frac{1}{R} \frac{\partial^4 \Phi_0}{\partial \eta^4} - \frac{\partial}{\partial t} \frac{\partial^2 \Phi_0}{\partial \eta^2} = -i\,k_0 \frac{\partial^2}{\partial \eta^2} \sum_{n=1}^{\infty} n \left[\bar{\Phi}_n \frac{\partial \Phi_n}{\partial \eta} - \Phi_n \frac{\partial \bar{\Phi}_n}{\partial \eta} \right] \tag{9.2.13}$$

It is seen that Eq. (9.2.13) can be integrated once, and even twice. However, it is sufficient to integrate once, since it is the first derivative of Φ_0 with respect to η which occurs in the system of equations and has a physical significance (that of a fluid velocity).

The constant of integration may be put equal zero if one introduces an *additional* condition, stating that the pressure gradient in Poiseuille flow is kept constant and independent of the perturbations (for demonstration see for instance Watson (1960)). Under this hypothesis we have for $\partial \Phi_0/\partial \eta$ the equation

$$\left[\frac{1}{R} \frac{\partial^2}{\partial \eta^2} - \frac{\partial}{\partial t}\right] \frac{\partial \Phi_0}{\partial \eta} = -i\,k_0 \frac{\partial}{\partial \eta} \sum_{n=1}^{\infty} n \left[\bar{\Phi}_n \frac{\partial \Phi_n}{\partial \eta} - \Phi_n \frac{\partial \bar{\Phi}_n}{\partial \eta} \right] \tag{9.2.14}$$

with boundary conditions (9.2.12).

9.3 Linearized stability theory

In the linearized theory the perturbations are assumed infinitesimally small and the non-linear terms $F_m^{(1)} + F_m^{(2)}$ are consequently neglected. Since in this approximation the Fourier components of different wavelength do not interact with each other, in keeping with the notation that we have used, we can consider the component Φ_1 and let k_0 take all positive values k. Moreover, the linearized equation admits solution of the form

$$\Phi_1(\eta, t) = e^{-\mu^{(k)} t} \varphi^{(k)}(\eta) \tag{9.3.1}$$

which lead, for the function $\varphi^{(k)}$, to the equation

$$\frac{1}{R}\left(\frac{\partial^2}{\partial\eta^2} - k^2\right)^2 \varphi^{(k)} + i k\left[\frac{d\varphi_0}{d\eta}\left(\frac{d^2}{d\eta^2} - k^2\right)\varphi^{(k)} - \frac{d^3\varphi_0}{d\eta^3}\varphi^{(k)}\right] + \\ + \mu^{(k)}\left(\frac{d^2}{d\eta^2} - k^2\right)\varphi^{(k)} = 0 \tag{9.3.2}$$

Eq. (9.3.2) is usually called the Orr-Sommerfeld equation, and is fundamental in many problems of hydrodynamic stability.

In the development of the general theory of the preceding chapters we have always assumed that the linearized stability problem was completely solved. Faced with the particular problem of the Poiseuille flow we have to admit that this is not entirely so.

Furthermore, the adjoint linearized problem, whose solutions are needed in the theory, has not yet been seriously studied. On the other hand, the theory that we have developed is based on certain hypotheses concerning the outcome of the linear theory and in every application one has to investigate whether these hypotheses are indeed verified. It is for these reasons that we shall briefly consider here the methods of solution of the Orr-Sommerfeld equation which have been developed in the past and summarize the results which have been obtained. The solution of the adjoint problem will be considered in section 9.4. As a result of that discussion we shall see more clearly to what extent it is possible to proceed with the application of the non-linear theory on the basis of the existing knowledge of the solutions of the linearized problem, and what elements of the linearized analysis, needed for the non-linear theory, are yet to be established.

The Orr-Sommerfeld equation is an ordinary differential equation of the fourth order. If we denote by $\varphi_1, \varphi_2, \varphi_3, \varphi_4$ the four independent solutions of this equation, the general solution can be represented by

$$\varphi^{(k)} = A_1\varphi_1 + A_2\varphi_2 + A_3\varphi_3 + A_4\varphi_4 \tag{9.3.3}$$

Application of the boundary conditions (9.2.11) leads to four linear homogeneous equations for the constants A_1, A_2, A_3, A_4, of which non-trivial solutions exist only if a certain determinant vanishes. This provides a relation between the parameters R, k and $\mu^{(k)}$. Now, the Reynolds number R is a given quantity for the flow under consideration, while the parameter k defines the wavelength of the perturbation with respect to which the stability is being studied. The determinant thus provides an equation determining the eigenvalues $\mu^{(k)}$ in terms of the parameters R and k.

To obtain a solution of the problem along these lines the explicit representation of $\varphi_1, \varphi_2, \varphi_3, \varphi_4$, is needed. However, guided by the experimental observation that instability of fluid motion usually occurs at large values of the Reynolds number, we may content ourselves with asymptotic expansions of the independent solutions, valid for $R \to \infty$. For this purpose it is convenient to change the notation in the Orr-Sommerfeld equation slightly. We denote the fluid velocities in the undisturbed Poiseuille flow by $\bar{u} = d\varphi_0/d\eta$ and write

$$\bar{u}'' = \frac{d^3\varphi_0}{d\eta^3}\,; \qquad \mu^{(k)} = -i\,k\,c^{(k)} \tag{9.3.4}$$

We then have

$$\frac{1}{k\,R}\left(\frac{d^2}{d\eta^2} - k^2\right)^2 \varphi^{(k)} + i\left\{[\bar{u} - c^{(k)}]\left(\frac{d^2}{d\eta^2} - k^2\right)\varphi^{(k)} - \bar{u}''\,\varphi^{(k)}\right\} = 0 \tag{9.3.5}$$

This suggests an asymptotic expansion

$$\varphi_{1.2} = \sum_{n=0}^{\infty} \frac{\psi_n}{(k\,R)^n} \tag{9.3.6}$$

The first approximation ψ_0 is easily seen to be a solution of the second order differential equation to which Eq. (9.3.5) reduces if we formally put $R \to \infty$. The two independent solutions $\varphi_{1.2}$ are therefore called the 'non-viscous solutions'.

The two remaining independent solutions can be found by the transformation

$$\varphi(\eta) = e^{\sqrt{kR}\,Q(\eta)} \tag{9.3.7}$$

that is, by the so-called W.K.B.J.-method. The resulting (non-linear) equation for Q admits in fact an asymptotic expansion of the form

$$Q = \sum_{n=0}^{\infty} \frac{Q_n}{(k\,R)^{\frac{1}{2}n}} \tag{9.3.8}$$

and one readily finds that

$$Q_0 = \mp \int^{\eta} \sqrt{-i(\bar{u} - c^{(k)})}\,d\eta'\,; \qquad Q_1 = -\frac{5}{4}\,l\,n[\bar{u} - c^{(k)}] \tag{9.3.9}$$

This leads to two independent 'viscous' solutions

$$\varphi_{3.4} = [\bar{u} - c^{(k)}]^{-\frac{5}{4}}\,e^{\mp\sqrt{kR}\int^{\eta}\sqrt{-i(\bar{u}-c^{(k)})}\,d\eta'}\{1 + 0(R^{-\frac{1}{2}})\} \tag{9.3.10}$$

With the construction of the four independent solutions in the form given above the true difficulties of the problem only begin to manifest themselves. This may be clearly seen in the case of neutral stability, where $\mu^{(k)}$ is purely imaginary, and thus $c^{(k)}$ is real, so that points

$\eta = \eta_c$ can arise such that $\bar{u} - c^{(k)} = 0$. In the vicinity of those' critical points' the asymptotic developments (9.3.6) and (9.3.10) become meaningless and other representations of the independent solutions must be sought. They can be obtained as 'inner asymptotic expansions', by introducing into Eq. (9.3.5) the transformation

$$\bar{\eta} = (k\,R)^{1/3}\,(\eta - \eta_c) \tag{9.3.11}$$

and seeking expansions for $R \to \infty$, $\bar{\eta}$ being now considered fixed. Finally, the matching of the 'inner' and 'outer' expansions must be accomplished. The reader is refered to C. C. LIN's book on hydrodynamic stability for the detailed description of the analysis and a discussion of the numerous delicate mathematical questions which have to be settled before the solution of the problem is obtained. For our purpose it is sufficiant to state that on the basis of these asymptotic methods the critical curve for neutral stability has been calculated. In general trend that curve agrees with the hypotheses formulated in section 6.3 and summarized in fig. 6.

We note in passing that it is here that the problem of Couette flow between two parallel walls must be disregarded as an application of the non-linear theory of the preceding chapters. In mathematical formulation the stability problem of Couette flow belongs to the class of problems defined in chapter 6. However, chapters 7 and 8 are concerned with weak stability or instability, defined in terms of points in the $(k - R)$ plane situated in the vicinity of the critical curve, while the linear theory of stability in Couette flow does not reveal the existence of such a curve.

The classical stability theory of Poiseuille flow achieves the determination of the critical curve, but to carry out the non-linear analysis of the preceding chapters we need further results. In principle, we require the knowledge of all eigenvalues $\mu_n^{(k)}$ and eigenfunctions $\varphi_n^{(k)}$ for every value of k and R. For the formal application we must at least convince ourselves that all the hypotheses underlying the non-linear theory are verified. For instance, our developments were based on the fundamental hypothesis that for every value of k and R there exists a denumerably infinite sequence of eigenvalues $\mu_n^{(k)}$ with the property

$$Re\,(\mu_{n+1}^{(k)}) > Re\,(\mu_n^{(k)}) \quad \text{for } k \text{ fixed and } n = 0, 1, 2, \ldots \tag{9.3.12}$$

The knowledge available on the solutions of the Orr-Sommerfeld equation is unfortunately not adequate to permit an investigation of the validity of the hypothesis. We are thus forced to proceed on a somewhat speculative basis, admitting without proof that relation (9.3.12) holds in the case of Poiseuille flow. However, to make that hypothesis more plausible we present a few arguments in its support.

We remark first, that if an infinite denumerable sequence of eigenvalues exists in the case of Poiseuille flow, then condition (9.3.12) must necessarily be satisfied, since otherwise there would be no critical curve. Alternatively, we can say that the classical calculation of the critical curve is based on the implicit hypothesis, that if other eigenvalues then the ones considered exist, they all have positive real parts.

The question of the number of eigenvalues of the Orr-Sommerfeld equation has received little attention in the past. Mention should be made of the work of MORAWETZ (1952); more explicit results are contained in CORCOS' and SELLARS' (1959) study of the stability of Poiseuille flow in a circular pipe.

The governing equation in that case is of course different from the equation of the two-dimensional problem, but it still is of the Orr-Sommerfeld type. Corcos and Sellars find a finite number of eigenvalues for every fixed value of R and k; they conclude and emphasize that the number of eigenvalues for the problem in question *is* finite, but this conclusion does not seem justified. In fact, the number of eigenvalues is found to be increasing with increasing R, while the calculations have been made, as usual, with only the first terms of the asymptotic expansions of the independent solutions for $R \to \infty$. The proper interpretation of the result should therefore be as follows: in the first asymptotic approximation for $R \to \infty$ the number of eigenvalues is found to be finite for every fixed value of R and increasing without limit as $R \to \infty$. In those terms, Corcos' and Sellars' result does not exclude the existence of an infinite denumerable sequence of eigenvalues; on the contrary, it provides an argument in support of that hypothesis.

Returning to the Orr-Sommerfeld equation we note that in certain limiting cases explicit solutions can be obtained without difficulty. Such is the case for $k = 0$ and for $R = 0$. In this last case, for example, we have Eq. (6.5.5) discussed in section 6.5. Imposing the boundary conditions, one finds an infinite denumerable sequence of eigenvalues, satisfying relation (9.3.12). Furthermore, by a perturbation technique the results can be extended to small values of R, and to small values of k. The validity of the hypothesis for the neighbourhood of $k = 0$, and for the neighbourhood of $R = 0$ is thus established. The necessary calculations are simple and are therefore omitted here.

The manner in which an infinite denumerable sequence of eigenvalues arises in the general case may be visualized by a different approach to the Orr-Sommerfeld equation. Instead of constructing the four independent solutions one can consider the eigenfunctions directly and expand them in series of suitable orthogonal functions which satisfy the boundary conditions of the problem.

The simplest system of orthogonal functions suitable for the present purpose, arises from the solutions of the equation

$$\frac{d^4 \psi}{d\eta^4} - \alpha^4 \psi = 0 \tag{9.3.13}$$

In order to keep within the established notation we change the interval of interest and consider $-\frac{1}{2} \leqq \eta \leqq +\frac{1}{2}$ instead of $0 \leqq \eta \leqq 1$. We impose the boundary conditions

$$(\psi)_{\eta=\frac{1}{2}} = (\psi)_{\eta=-\frac{1}{2}} = \left(\frac{d\psi}{d\eta}\right)_{\eta=\frac{1}{2}} = \left(\frac{d\psi}{d\eta}\right)_{\eta=-\frac{1}{2}} = 0 \tag{9.3.14}$$

The solutions can now be grouped in two classes: the symmetrical eigenfunctions

$$\psi_m = \frac{\cosh \alpha_m \eta}{\cosh \frac{1}{2} \alpha_m} - \frac{\cos \alpha_m \eta}{\cos \frac{1}{2} \alpha_m} \tag{9.3.15}$$

and the antisymmetrical ones

$$\psi_m = \frac{\sin \alpha_m \eta}{\sinh \frac{1}{2} \alpha_m} - \frac{\sin \alpha_m \eta}{\sin \frac{1}{2} \alpha_m} \tag{9.3.16}$$

The eigenvalues of the problem for ψ, for both Eq. (9.3.15) and Eq. (9.3.16) can be found for instance in CHANDRASEKHAR (1961). One easily establishes the orthogonality of the functions ψ_m: from quite general considerations (COURANT-HILBERT) one can prove an expansion theorem which states that any continuous function with continuous derivatives up to third order and a piecewise continuous fourth derivative, satisfying the boundary conditions (9.3.14), can be expanded in absolutely and uniformly convergent series of ψ_m.

Any solution of the Orr-Sommerfeld equation which satisfies those continuity conditions and the boundary conditions (9.3.14), can hence be represented by the series

$$\varphi^{(k)} = \sum_{n=0}^{\infty} a_n \psi_n \tag{9.3.17}$$

Multiplying the Orr-Sommerfeld equation by the function $\psi_m(\eta)$ and integrating over the interval one finds

$$\sum_{n=1}^{\infty} X_n^{(m)} a_n = 0 \tag{9.3.18}$$

where $X_n^{(m)}$ are coefficients whose explicit form is easily established in terms of some standard integrals, most of which can be found tabulated in CHANDRASEKHAR (1961). The coefficients $X_n^{(m)}$ are linear functions of $\mu^{(k)}$.

If we now consider an approximation of order M, that is, if in the series (9.3.17) all terms for which $n > M$ are neglected, then in Eq. (9.3.18) we have a system of M equations for M unknowns a_n. However, the system is homogeneous so that non-trivial solutions only exist if the determinant of the coefficients $X_n^{(m)}$ vanishes. Since $X_n^{(m)}$ are linear functions of $\mu^{(k)}$, the determinant, when developed leads to an algebraic equation of M-th degree in $\mu^{(k)}$. Consequently, we shall obtain M distinct values of the parameter $\mu^{(k)}$, for which the problem admits non-trivial solutions. Finally, as M increases, the number of distinct eigenvalues $\mu^{(k)}$ increases without limit.

The procedure of expansion in orthogonal functions has been little used in the past in connection with the Orr-Sommerfeld equation. In comparison with the asymptotic method outlined in the begining of this section it has the disadvantage of heavier numerical calculations. But it presents an interesting alternative and, from the present point of view, it has the advantage of showing the mechanism by which an infinite sequence of eigenvalues for the Orr-Sommerfeld equation may arise. Of course, expansion in functions other then those defined in Eqs. (9.3.15), (9.3.16) can also be used, such as for instance the functions defined in Eq. (6.5.6). In that case the functions are no longer mutually orthogonal in the ordinary sense, but possess the orthogonality property (6.5.8), which seems even better adapted to the present problem.

For the case of Poiseuille flow the method described above has been developed to some extent by MAUSS (1963), where details concerning the coefficients $X_n^{(m)}$, etc. can be found. The problem of stability of Couette flow has been studied by the orthogonal functions method by GALAGHER and MERCER (1962).

9.4 The adjoint linearized problem

We shall discuss now some of the more important properties of the solution of the adjoint problem. In mathematical complexity this problem is comparable to that of the preceding section; it is therefore impossible to present a complete satisfactory analysis in the context of this study. In what follows we concentrate on bringing out the relations which exist between the solutions of the Orr-Sommerfeld equation and those of its adjoint; in that process most of the necessary demonstrations are only briefly outlined.

Following section 6.5, we write the Orr-Sommerfeld equation in the form

$$[\mathfrak{L}_k - i\,k\,c^{(k)}\,S_k]\,\varphi^{(k)} = 0 \tag{9.4.1}$$

$$\mathfrak{L}_k = \left(\frac{d^2}{d\eta^2} - k^2\right)^2 + i\,k\,R\left[\bar{u}\left(\frac{d^2}{d\eta^2} - k^2\right) - \bar{u}''\right] \tag{9.4.2}$$

$$S_k = \frac{d^2}{d\eta^2} - k^2; \qquad \bar{u}' = \frac{d\bar{u}}{d\eta}; \qquad \bar{u}'' = \frac{d^2\bar{u}}{d\eta^2} \tag{9.4.3}$$

Since the operator S_k is self-adjoint, we only need to define the adjoint of the operator $\mathfrak{L}_k$. This adjoint operator is given by

$$\tilde{\mathfrak{L}}_k = \left(\frac{d^2}{d\eta^2} - k^2\right)^2 + i\,k\,R\left[\bar{u}\left(\frac{d^2}{d\eta^2} - k^2\right) + 2\,\bar{u}'\frac{d}{d\eta}\right] \tag{9.4.4}$$

In fact, a simple calculation shows that

$$\tilde{\varphi}^{(k)}\,\mathfrak{L}_k[\varphi^{(k)}] - \varphi^{(k)}\,\tilde{\mathfrak{L}}_k[\tilde{\varphi}^{(k)}] = \frac{d}{d\eta}\left\{\tilde{\varphi}^{(k)}\frac{d^3\varphi^{(k)}}{d\eta^3} - \varphi^{(k)}\frac{d^3\tilde{\varphi}^{(k)}}{d\eta^3} - \right.$$
$$- \frac{d\tilde{\varphi}^{(k)}}{d\eta}\frac{d^2\varphi^{(k)}}{d\eta^2} + \frac{d\varphi^{(k)}}{d\eta}\frac{d^2\tilde{\varphi}^{(k)}}{d\eta^2} - 2k^2\left(\tilde{\varphi}^{(k)}\frac{d\varphi^{(k)}}{d\eta} - \varphi^{(k)}\frac{d\tilde{\varphi}^{(k)}}{d\eta}\right) +$$
$$\left. + i\,k\,R\left[\bar{u}\,\tilde{\varphi}^{(k)}\frac{d\varphi^{(k)}}{d\eta} - \varphi^{(k)}\frac{d u_0\,\tilde{\varphi}^{(k)}}{d\eta}\right]\right\} \tag{9.4.5}$$

Moreover, the bilinear concomitant vanishes at the boundary points if $\tilde{\varphi}^{(k)}$ is subject to the same boundary conditions as $\varphi^{(k)}$.

Summarizing, we define the adjoint problem by

$$\left(\frac{d^2}{d\eta^2} - k^2\right)^2\tilde{\varphi}^{(k)} + i\,k\,R\left\{(\bar{u} - \tilde{c}^{(k)})\left(\frac{d^2}{d\eta^2} - k^2\right)\tilde{\varphi}^{(k)} + 2\,\bar{u}'\frac{d\tilde{\varphi}^{(k)}}{d\eta}\right\} = 0 \tag{9.4.6}$$

with boundary conditions

$$\tilde{\varphi}^{(k)}(0) = \tilde{\varphi}^{(k)}(1) = \tilde{\varphi}^{(k)}_\eta(0) = \tilde{\varphi}^{(k)}_\eta(1) = 0 \tag{9.4.7}$$

From equations (9.3.5) and (9.4.6) we immediately obtain the orthogonality relation between the eigenfunctions of the Orr-Sommerfeld equation and its adjoint

$$[\tilde{c}^{(k)} - c^{(k)}]\int_0^1 \tilde{\varphi}^{(k)}\left(\frac{d^2}{d\eta^2} - k^2\right)\varphi^{(k)}\,d\eta = 0 \tag{9.4.8}$$

In order to determine the adjoint eigenfunctions explicitly we can proceed as is section 9.3 and first construct the asymptotic expansions for $R \to \infty$ of the four independent solutions of Eq. (9.4.6). Thus we introduce the asymptotic series

$$\tilde{\varphi}^{(k)}_{1,2} = \sum_{n=0}^{\infty}\frac{\tilde{\psi}_n}{(k\,R)^n} \tag{9.4.9}$$

The first approximation $\tilde{\psi}_0$ is then the solution of the equation

$$[\bar{u} - \tilde{c}^{(k)}]\left(\frac{d^2}{d\eta^2} - k^2\right)\tilde{\psi}_0 + 2\,\bar{u}'\frac{d\tilde{\psi}_0}{d\eta} = 0 \tag{9.4.10}$$

which can be written in the form

$$\left(\frac{d^2}{d\eta^2} - k^2\right)[(\bar{u} - \tilde{c}^{(k)})\,\tilde{\psi}_0] - \bar{u}''\,\tilde{\psi}_0 = 0 \tag{9.4.11}$$

Comparing with Eqs. (9.3.5), (9.3.6) we see that the quantity $[\bar{u} - \tilde{c}^{(k)}]\tilde{\varphi}_0$ is a solution of the equation which determined ψ_0. In other words, in the first asymptotic approximation the 'non-viscous' independent solutions of the Orr-Sommerfeld equation are related to the 'non-viscous' independent solutions of the adjoint equation by the formula

$$\tilde{\varphi}^{(k)}_{1.2} = \frac{\varphi^{(k)}_{1.2}}{\bar{u} - \tilde{c}^{(k)}} \tag{9.4.12}$$

where $\varphi^{(k)}_{1.2}$ is to be considered formally as a function of $\tilde{c}^{(k)}$ instead of $c^{(k)}$.

Turning now to the 'viscous' solutions, we introduce in Eq. (9.4.6) the transformation

$$\tilde{\varphi}^{(k)} = e^{\sqrt{kR}\,\tilde{Q}} \tag{9.4.13}$$

One easily finds the asymptotic expansion

$$\tilde{Q} = \sum_{n=0}^{\infty} \frac{\tilde{Q}_n}{(kR)^{\frac{1}{2}n}} \tag{9.4.14}$$

$$\tilde{Q}_0 = \mp \int^{\eta} \sqrt{-i[\bar{u} - \tilde{c}^{(k)}]}\, d\eta'; \qquad \tilde{Q}_1 = -\tfrac{1}{4} \ln [\bar{u} - \tilde{c}^{(k)}] \tag{9.4.15}$$

Comparing with Eq. (9.3.10) we find the relation between the 'viscous' independent solution of the Orr-Sommerfeld equation and its adjoint

$$\tilde{\varphi}^{(k)}_{3.4} = [\bar{u} - \tilde{c}^{(k)}]\, \varphi^{(k)}_{3,4} \tag{9.4.16}$$

where again $\varphi^{(k)}_{3.4}$ is to be considered formally as function of $\tilde{c}^{(k)}$, instead of $c^{(k)}$.

Relations (9.4.12) and (9.4.16) are particularly useful when we study the determinant which arises from the application of the boundary conditions (9.4.7) to the general solution of the adjoint equation (9.4.6). When that determinant is expanded for $R \to \infty$ in a manner explained in Lin (1955) chapter 3.4, the first approximation proves to be formally identical with the first approximation from which $c^{(k)}$ is calculated in the Orr-Sommerfeld problem. Thus an illustration is obtained of a general property which can be stated as follows:

To every eigenvalue $c^{(k)}_n$ and eigenfunction $\varphi^{(k)}_n$ of the Orr-Sommerfeld problem there corresponds an adjoint eigenfunction $\tilde{\varphi}^{(k)}_n$ with eigenvalue $\tilde{c}^{(k)}_n = c^{(k)}_n$ of the adjoint problem.

The orthogonality property (9.4.8) can finally, be written after normalization in the form

$$\int_0^1 \left[\frac{d\tilde{\varphi}^{(k)}_n}{d\eta} \frac{d\varphi^{(k)}_m}{d\eta} + k^2 \tilde{\varphi}^{(k)}_n \varphi^{(k)}_m \right] d\eta = \begin{cases} 1 \text{ if } n = m \\ 0 \text{ if } n \neq m \end{cases} \tag{9.4.17}$$

The analysis outlined above holds for arbitrary functions $\bar{u}(\eta)$. It applies to Poiseuille flow, but also to Couette flow and other more general velocity distributions $\bar{u}(\eta)$. In the particular case of Poiseuille flow, that is for $\bar{u}(\eta) = \eta(1-\eta)$, another relation between the solutions of the Orr-Sommerfeld equation and its adjoint has been discovered by STUART (1960b). Stuart observes that if $\varphi^{(k)}$ is a solution of the Orr-Sommerfeld equation, the function

$$f(\eta) = \frac{d^2 \varphi^{(k)}}{d\eta^2} - k^2 \varphi^{(k)} \tag{9.4.18}$$

will satisfy the adjoint equation (9.4.6) in the case of Poiseuille flow. Thus if we take the general solution of the Orr-Sommerfeld equation for $\varphi^{(k)}$, relation (9.4.18) gives the general solution of the adjoint equation. We remark, however, that if for $\varphi^{(k)}$ we take an eigenfunction of the Orr-Sommerfeld equation, we do not obtain the adjoint eigenfunction by Eq. (9.4.18), because the boundary conditions are not satisfied.

9.5 Periodic solutions

We now proceed to the formal application of the theory of chapter 7 to the problem of Poiseuille flow. We consider weak stability or instability, that is we assume that in the $k-R$ plane the point k_0, R is close to the critical curve. To simplify the notation we write again $\varphi_n^{(m k_0)} = \varphi_n^{(m)}$; $\tilde{\varphi}_n^{(m k_0)} = \tilde{\varphi}_n^{(m)}$; $\mu_n^{(m k_0)} = \mu_n^{(m)}$; etc.

In Eqs. (9.2.7), (9.2.10), (9.2.14) we introduce the transformation

$$\Phi_m = \varepsilon\, \delta_m\, \psi_m \tag{9.5.1}$$

$$\delta_m = \varepsilon^{|m-1|}; \qquad \varepsilon^2 = |\tau_0|; \qquad \tau_0 = Re[\mu_0^{(1)}] \tag{9.5.2}$$

The asymptotic expansion for $\varepsilon \to 0$ now yields the following set of equations

$$\left[\frac{1}{R}\frac{\partial^2}{\partial\eta^2} - \frac{\partial}{\partial t}\right]\frac{\partial\psi_0}{\partial\eta} = -i k_0 \frac{\partial}{\partial\eta}\left[\bar{\psi}_1 \frac{\partial\psi_1}{\partial\eta} - \psi_1 \frac{\partial\bar{\psi}_1}{\partial\eta}\right] + O(\varepsilon^2) \tag{9.5.3}$$

$$\begin{aligned}\left[\mathfrak{L}_1 - \frac{\partial}{\partial t} S_1\right]\psi_1 = -\varepsilon^2 i k_0 \left[\frac{\partial\psi_0}{\partial\eta} S_1(\psi_1) - \frac{\partial^3\psi_0}{\partial\eta^3}\psi_1 + \bar{\psi}_1 \frac{\partial}{\partial\eta} S_1(\psi_2) -\right.\\ \left. - 2\psi_2 \frac{\partial}{\partial\eta} S_1(\bar{\psi}_1) + 2\frac{\partial\bar{\psi}_1}{\partial\eta} S_2(\psi_2) - \frac{\partial\psi_2}{\partial\eta} S_1(\bar{\psi}_1)\right] + O(\varepsilon^4)\end{aligned} \tag{9.5.4}$$

$$\left[\mathfrak{L}_2 - \frac{\partial}{\partial t} S_2\right]\psi_2 = i k_0 \left[\psi_1 \frac{\partial^3\psi_1}{\partial\eta^3} - \frac{\partial\psi_1}{\partial\eta}\frac{\partial^2\psi_1}{\partial\eta^2}\right] + O(\varepsilon^2) \tag{9.5.5}$$

We recall the definition of the operators

$$S_m = \frac{\partial^2}{\partial\eta^2} - m^2 k_0^2 \tag{9.5.6}$$

$$\mathfrak{L}_m = \frac{1}{R} S_m^2 + i m k_0 \left[\frac{d\varphi_0}{d\eta} S_m - \frac{d^3\varphi_0}{d\eta^3}\right] \tag{9.5.7}$$

We introduce for ψ_1 the eigenfunction expansion

$$\psi_1(\eta, t) = \sum_{n=0}^{\infty} A_n^{(1)}(t)\, \varphi_n^{(1)}(\eta) \tag{9.5.8}$$

From the theory of section 7.3 we know that

$$A_n^{(1)} = 0(\varepsilon^2) \quad \text{for} \quad n > 0 \tag{9.5.9}$$

Utilizing these results in Eqs. (9.5.3), (9.5.5), and following the analysis of section 7.4, the forced solutions for ψ_0 and ψ_2 are obtained in the form

$$\psi_0(\eta, t) = |A_0^{(1)}|^2 g_0(\eta) + 0(\varepsilon^2) \tag{9.5.10}$$

$$\psi_2(\eta, t) = [A_0^{(1)}]^2 g_2(\eta) + 0(\varepsilon^2) \tag{9.5.11}$$

where the function g_2 is defined as the solution of the equation

$$[\mathfrak{L}_2 + 2i\, Im(\mu_0^{(1)})\, S_2]\, g_2 = i k_0 \left[\varphi_0^{(1)} \frac{d^3 \varphi_0^{(1)}}{d\eta^3} - \frac{d\varphi_0^{(1)}}{d\eta} \frac{d^2 \varphi_0^{(1)}}{d\eta^2}\right] \tag{9.5.12}$$

satisfying the boundary conditions

$$g_2(0) = g_2(1) = 0; \quad \left(\frac{d g_2}{d\eta}\right)_{\eta=0} = \left(\frac{d g_2}{d\eta}\right)_{\eta=1} = 0 \tag{9.5.13}$$

As we have seen in chapter 7, an eigenfunction expansion for the function g_2 can easily be given.

The function g_0 can be defined in a similar way; however, in the present problem only the function $\partial\psi_0/\partial\eta$, and hence $d g_0/d\eta$, is of interest. Following section 7.4 again we find

$$\frac{1}{R} \frac{d^2}{d\eta^2}\left(\frac{d g_0}{d\eta}\right) = -i k_0 \frac{d}{d\eta}\left[\bar{\varphi}_0^{(1)} \frac{d\varphi_0^{(1)}}{d\eta} - \varphi_0^{(1)} \frac{d\bar{\varphi}_0^{(1)}}{d\eta}\right] \tag{9.5.14}$$

The integration can be performed immediately and taking into account the boundary conditions (9.2.12) we obtain

$$\frac{d g_0}{d\eta} = -i k_0 R \left\{\int_0^{\eta} \left[\bar{\varphi}_0^{(1)} \frac{d\varphi_0^{(1)}}{d\eta'} - \varphi_0^{(1)} \frac{d\bar{\varphi}_0^{(1)}}{d\eta'}\right] d\eta' - \right.$$

$$\left. - \eta \int_0^{1} \left[\bar{\varphi}_0^{(1)} \frac{d\varphi_0^{(1)}}{d\eta'} - \varphi_0^{(1)} \frac{d\bar{\varphi}_0^{(1)}}{d\eta'}\right] d\eta' \right\} \tag{9.5.15}$$

Turning to the component ψ_1 we utilize the results (9.5.10), (9.5.11) in Eq. (9.5.4) and find

$$\left[\mathfrak{L}_1 - \frac{\partial}{\partial t} S_1\right] \psi_1 = -\varepsilon^2 |A_0^{(1)}|^2 A_0^{(1)} f_1(\eta) + 0(\varepsilon^4) \tag{9.5.16}$$

$$f_1(\eta) = -i k_0 \left\{\frac{d g_0}{d\eta} S_1[\varphi_0^{(1)}] - \frac{d^3 g_0}{d\eta^3} \varphi_0^{(1)} + \bar{\varphi}_0^{(1)} \frac{d}{d\eta} S_2(g_2) - \right.$$

$$\left. - 2 g_2 \frac{d}{d\eta} S_1[\bar{\varphi}_0^{(1)}] + 2 \frac{d\bar{\varphi}_0^{(1)}}{d\eta} S_2(g_2) - \frac{d g_2}{d\eta} S_1[\varphi_0^{(1)}]\right\} \tag{9.5.17}$$

Consequently, by application of the eigenfunction expansion (9.5.8) we have

$$\frac{d A_0^{(1)}}{dt} + \mu_0^{(1)} A_0^{(1)} = -\varepsilon^2 A_0^{(1)} |A_0^{(1)}|^2 \beta_0 + 0(\varepsilon^4) \qquad (9.5.18)$$

$$\beta_0 = \int_0^1 f_1(\eta)\, \tilde{\varphi}_0^{(1)}(\eta)\, d\eta \qquad (9.5.19)$$

Eq. (9.5.18) admits solutions harmonic in time

$$\overset{\infty}{A_0^{(1)}}(t) = e^{i\omega t} a_0^{(1)} \qquad (9.5.20)$$

$$|a_0^{(1)}|^2 = -\frac{\tau_0}{|\tau_0|} [Re(\beta_0)]^{-1} \qquad (9.5.21)$$

$$\omega = -Im[\mu_0^{(1)}] - |\tau_0|\, Im(\beta_0)\, |a_0^{(1)}|^2 + 0(\varepsilon^4) \qquad (9.5.22)$$

On the basis of the discussion of section 7.6 we know that if $Re(\beta_0) > 0$ the harmonic solution exists in the case of weak instability, and

$$\lim_{t\to\infty} |A_0^{(1)}|^2 = |a_0^{(1)}|^2 \qquad (9.5.23)$$

If $Re(\beta_0) < 0$, however, the harmonic solution exists in the case of weak stability and must be interpreted as a critical value for instability.

We remark finally that the preceding results are valid subject to the condition

$$Re[\beta_0] = 0(1) \quad \text{for} \quad \varepsilon \to 0 \qquad (9.5.24)$$

The fundamental result (9.5.18), —with the constant β_0 defined by Eqs. (9.5.19), (9.5.17) and the functions g_0, g_2 defined by Eqs. (9.5.12), (9.5.14)—, is formally equivalent to the results of the analysis of STUART (1960b).

9.6 Discussion of the results

We have seen that application of the theory of chapter 7 leads to a confirmation of STUART's (1960b) pioneering work on periodic solutions in Poiseuille flow. Furthermore, if $Re(\beta_0) > 0$, we can proceed to interpret those solutions in the light of the analysis of chapter 8.

We return to fig. 10, which in basic aspects can be considered to represent the critical curve for Poiseuille flow, at least in the vicinity of $R = R_{cr}$ and $k = k_{cr}$. We consider a fixed Reynolds number $R > R_{cr}$, the quantity $|R - R_{cr}|$ being small.

If now $R = R_0(k)$ is the equation of the critical curve, we define a positive constant γ_2 by the relation

$$\gamma_2 = \frac{1}{2}\left(\frac{d^2 R_0}{dk^2}\right)_{k=k_{cr}} \qquad (9.6.1)$$

In the case $Re(\beta_0) > 0$, periodic solutions with period $2\pi/k_0$ exist for all values of k_0 satisfying the relation

$$|k_0 - k_{cr}| < \sqrt{\frac{R - R_{cr}}{\gamma_2}} \tag{9.6.2}$$

The principal result of chapter 8 states that in the interval (9.6.2) a subinterval

$$|k_0 - k_{cr}| < \sqrt{\frac{R - R_{cr}}{3\gamma_2}} \tag{9.6.3}$$

should be distinguished; if k_0 satisfies condition (9.6.3) the corresponding periodic solution is stable with respect to small aperiodic perturbations.

Consequently, we can conclude that in the problem of stability of Poiseuille flow for $R > R_{cr}$, the stability with respect to arbitrary (and hence aperiodic) small perturbations being considered, the periodic solutions that satisfy condition (9.6.3) represent a possible limit of the perturbations for $t \to \infty$. If a two-dimensional experiment were possible, one of those stable periodic solutions could ultimately be observed. (For further discussion the reader is referred back to section 8.12.)

The above interpretation applies to the case $Re(\beta_0) > 0$, and in the final analysis no conclusion about the behaviour of perturbations in Poiseuille flow can be reached without, at least, the knowledge of the sign of $Re(\beta_0)$. It appears that this knowledge can only be obtained from an actual computation of the constant β_0. We shall presently review the steps needed to perform such a calculation.

The eigenfunctions $\varphi_0^{(1)}$ and its complex conjugates $\bar{\varphi}_0^{(1)}$ can be considered to be known from the linearized theory in the form of asymptotic series for $R \to \infty$. The adjoint eigenfunctions can be obtained in similar form, as outlined in section 9.4.

The function $dg_0/d\eta$ was determined explicitly in Eq. (9.5.15). The missing piece of information, necessary for the calculation of β_0, is the function g_2. That function is defined as the solution of Eq. (9.5.12) subject to boundary condition (9.5.13). The eigenfunction expansion discussed in section 7.4 does not seem practical in the present case since, as we have seen, the eigenfunctions of the Orr-Sommerfeld equation are difficult to construct. The function g_2 must be determined by other methods, such as expansion in series of suitable orthogonal functions, or integration with the aid of the four independent solutions of the Orr-Sommerfeld equation, or, finally, direct numerical integration. In any case the difficulties are considerable and the determination of g_2 appears to be the most serious obstacle in the calculation of β_0. Moreover, even when g_2 is determined, the calculation of the integral which defines β_0 can still present considerable numerical labor.

The practical difficulties which arise in the application of the non-linear analysis to the problem of Poiseuille flow result, of course, from the complexity of the linearized theory. More generally it can be expected that non-linear analysis will lead to simple results as long as the linearized problem is relatively simple (see the example treated in section 7.7). Computational difficulties will however most certainly arise in every problem for which the linearized theory is already complicated. In such cases, of which the problem of Poiseuille flow is an example, the present non-linear theory succeeds in bringing the solution into view, but the importance of the numerical analysis which remains to be done should by no means be underestimated.

BIBLIOGRAPHY

I. Research on non-linear stability problems

BENNEY, D. J., and C. C. LIN: On the secondary motion induced by oscillations in shear flow. J. Phys. Fluids **3**, 656—657 (1960).

BENNEY, D. J.: A non-linear theory of oscillations in parallel flow. J. Fluid Mech. **10**, 209—236 (1961).

ECKHAUS, W.: Problèmes non linéaires dans la théorie de la stabilité. J. Mécanique I, 49—77 (1962).

—: Problèmes non linéaires de stabilité dans un espace à deux dimensions. Première partie: Solutions périodiques. J. Mécanique I, 413—438 (1962).

—: Problèmes non linéaires de stabilité dans un espace à deux dimensions. Deuxième partie: Stabilité des solutions périodiques. J. Mécanique II, 153—172 (1963).

GÖRTLER, H., and H. WITTING: Theorie der sekundären Instabilität der laminaren Grenzschichten. IUTAM Symposium Freiburg, Berlin/Göttingen/Heidelberg: Springer 1957, 110—126.

HEISENBERG, W.: Über Stabilität und Turbulenz von Flüssigkeitsströmen. Ann. Phys. Lpz. **74**, 577—627 (1924).

KUO, H. L.: Solution of the non-linear equations of cellular convection and heat transport. J. Fluid Mech. **10**, 611—634 (1961).

LANDAU, L. D.: C. R. Acad. Sci. URSS **44**, 311—314 (1944).

LIN, C. C.: On the instability of laminar flow and its transition to turbulence. IUTAM Symposium Freiburg, Berlin/Göttingen/Heidelberg: Springer 1957, 144—160.

MALKUS, W. V. R., and G. VERONIS: Finite amplitude cellular convection. J. Fluid Mech. **10**, 225—260 (1958).

MEKSYN, D., and J. T. STUART: Stability of viscous motion between parallel planes for finite disturbances. Proc. Roy Soc., A. Lond., **208**, 517—526 (1951).

NOETHER, F.: Das Turbulenzproblem. Z. angew. Math. Mech. I, 125—138 (1921).

SEGEL, A. L., and J. T. STUART: On the question of preferred mode in cellular termal convection. J. Fluid Mech. **13**, 289—306 (1962).

SEGEL, A. L.: The non-linear interaction of two disturbances in the termal convection problem. J. Fluid Mech. **14**, 97—114 (1962).

STRUMINSKI, W. W.: On the theory of aerodynamic stability. Doklady Akad. Nauk SSSR **151**, 1046—1049 (1963a).

—: On the non-linear theory of the development of perturbations in aerodynamics. Doklady Akad. Nauk SSSR **153**, 547—550 (1963b).

STUART, J. T.: On the non-linear mechanics of hydrodynamic stability. J. Fluid Mech. **4**, 1—21 (1958).

—: Non-linear effects in hydrodynamics stability. Proc. Xth Int. Congr. Appl. Mech. Stresa 63—97 (1960a).

—: On the non-linear mechanics of wave disturbances in stable and unstable parallel flows. Part. 1. The basic behaviour in plane Poiseuille flow. J. Fluid Mech. **9**, 353—370 (1960b).

—: On the three-dimensional non-linear effects in the stability of parallel flow. Adv. Aero. Sci. **3—4**, 121—142 (1961).

VELTE, W.: Stabilitätsverhalten und Verzweigung stationärer Lösungen der Navier-Stokesschen Gleichungen. Arch. Rat. Mech. Anal. **16**, 97—125 (1964).

WATSON, J.: On the non-linear mechanics of wave disturbances in stable and unstable parallel flow. Part 2. The development of a solution for plane Poiseuille flow and for plane Couette flow. J. Fluid Mech. **9**, 371—389 (1960).

—: On spatially-growing finite disturbances in plane Poiseuille flow. J. Fluid Mech. **14**, 211—221 (1962).

II. Other publications cited in the text

BATCHELOR, K. G.: The theory of homogeneous turbulence. Cambridge University Press (1956).

BENARD, M.: Les turbillions cellulaires dans une nappe liquide transportant de la chaleur par convection en régime permanant. Ann. Chim. Phys. **23**, 62—144 (1901).

BURGERS, J. M.: A mathematical model illustrating the theory of turbulence. Adv. Appl. Mech. **1**, 171—196 (1946).

CHANDRASEKHAR, S.: Hydrodynamic and hydromagnetic stability. Oxford University Press (1961).

CORCOS, G. M., and J. R. SELLARS: On the stability of fully developed flow in a pipe. J. Fluid Mech. **5**, 97—112 (1959).

COURANT, R., and D. HILBERT: Methods of mathematical physics. Intersciences Publishers Vol. I (1953), Vol. II (1962).

ECKHAUS, W.: Theory of flame-front stability. J. Fluid Mech. **10**, 80—100 (1961).

GALAGHER, A. P., and A. MERCER: On the behaviour of small disturbances in plane Couette flow. J. Fluid Mech. **13**, 91—100 (1962).

KODAIRA, K: On ordinary differential equations of any even order and the corresponding eigenfunction expansions. Am. Journ. of Math. **72**, 502—544 (1950).

LANDAU, L. D., and E. M. LIFSCHITZ: Fluid mechanics. Pergamon Press (1959).

LEITE, R. J.: An experimental investigation of the stability of Poiseuille flow. J. Fluid Mech. **5**, 81—96 (1959).

LIN, C. C.: The theory of hydrodynamic stability. Cambridge University Press (1955).

MARKSTEIN, G. H.: Experimental and theoretical studies of flame stability J. Aero Sci. **18** (1951).

MAUSS, J.: Etude de la stabilité de l'écoulement de Poiseuille par la méthode des fonctions orthogonales. Département de Mécanique, Paris (1964).

MORAWETZ, C. S.: The eigenvalues of some stability problems involving viscosity. J. Rat. Mech. Anal. **1**, 579—603 (1952).

MORSE, P. M., and H. FESHBACH: Methods of theoretical physics. McGraw-Hill (1953).

RAYLEIGH, Lord: On convection currents in a horizontal layer of fluid when the higher temperature is on the under side. Scient. Papers **6**, 432—466 (1916).

REYNOLDS, O.: The experimental investigation of the circumstances which determine whether the motion of water should be direct or sinuous. Trans. Roy. Soc. London A **174**, 935—982 (1883).

LA SALLE, J., and S. LEFSCHETZ: Stability by Liapunov's direct method. Academic Press (1961).

SATO, H., and K. KURIKI: The mechanism of transition in the wake of a thin plate placed parallel to the uniform flow. J. Fluid Mech. **11**, 321—352 (1961).

SCHUBAUER, G. B., and H. K. SKRAMSTAD: Laminar boundary layer oscillation and transition on a flat plate. J. Aero Sci. **14** (1947).

SCHUBAUER, G. B., and P. S. KLEBANOFF: Contribution on the mechanics of boundary layer transition. NACA Rep. 1289 (1956).

TAYLOR, G. I.: Stability of viscous liquid contained between two rotating cylinders. Phil. Trans. Roy. Soc. A. **223**, 289—343 (1923).

INDEX